IMAGES OF DEVELOPMENT

SUNY Series in Philosophy and Biology
David Edward Shaner, Editor

IMAGES OF DEVELOPMENT

Environmental Causes in Ontogeny

Cor van der Weele

WITHDRAWN

State University of New York Press

Published by
State University of New York Press, Albany

© 1999 State University of New York

For information, address State University of New York Press,
State University Plaza, Albany, New York, 12246

Production by Diane Ganeles
Marketing by Anne Valentine

Library of Congress Cataloging-in-Publication Data

Weele, Cor van der, 1954–
    Images of development : environmental causes in ontogeny / Cor van
der Weele.
        p.   cm. — (SUNY series in philosophy and biology)
    Includes bibliographical references and index.
    ISBN 0-7914-4045-1 (hardcover : alk. paper). — ISBN 0-7914-4046-X
(pbk. : alk. paper)
    1. Ontogeny.   2. Ecology.   3. Developmental biology.   I. Title.
II. Series.
QH491.W443   1999
591.3'8—dc21   98-14899
                                                                    CIP

10 9 8 7 6 5 4 3 2 1

The map butterfly of European map, *Araschnia levana*
Left: summer form, black with a white band
Right: spring form, orange and black

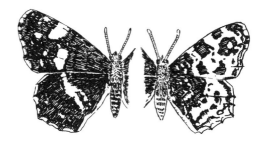

Contents

Contents

Illustrations

Foreword

Ontogeny belongs to biology. My subject is the study of ontogeny, and although I do consider what biologists have to say about ontogeny, this is not really a biology book. I try to take a few steps back from biology to discover patterns, trends, and omissions. Neither am I writing philosophy, not pure philosophy at any rate, because what I do is in the service of biological, not philosophical, problems. The location of the discussion is somewhere in-between biology and philosophy. This in-between area is not as well developed as the mother disciplines, but it is a precious and challenging domain; it is here that questions about such things as goals and presuppositions, alternative concepts, blind spots, and explanatory patterns in biology are at home and can get the attention they deserve. I bring up such questions in connection with a rather straightforward issue: the neglect of environmental influences that I observed in the study of embryological development in animals.

Some people may think that my approach is even a little too straightforward. Focusing attention on environmental influences in development might reinforce the tiresome dichotomy between genes and environment, they might argue; instead we should emphasize interaction and complexity. I do agree, as will be evident throughout, that an acknowledgment of causal interaction and complexity offers the right overall perspective. But within that perspective, I think that specific questions about environmental influences in the developmental network should be asked, in order to attain a balance of attention that has been lacking until now. With the growing recognition that environmental toxins can be potent disruptors of development, it becomes ever more clear that knowledge about both normal and abnormal environmental influences is of more than academic interest. Indeed, behind my questions lies the conviction that the study of ontogeny matters outside biology and outside philosophy.

The manuscript of this book was finished more than two years ago, in May 1995, in the form of a Ph.D. thesis. Patterns and trends, the issues that the book deals with, change at a much slower pace than scientific detail. Nevertheless, on this timescale too, things happen. I decided that adding a coda, in which I mention recent developments and comment on some aspects of the changing landscape, is a better way to show something of this dynamic than adapting the main text a little here and there. In the coda, I will mention the surprising new edition of Scott Gilbert's textbook *Developmental Biology,* among other things.

The land between biology and philosophy is not entirely unpopulated. Operating firmly between disciplines, there is the International Society for the History, Philosophy and Social Studies of Biology. Its conferences allowed me to meet many colleagues. Let me mention in particular Ron Amundson, Richard Burian, Linnda Caporael, Fred Gifford, Scott Gilbert, Brian Goodwin, Russell Gray, Paul Griffiths, Donna Haraway, Evelyn Fox Keller, Lenny Moss, Eva Neumann-Held, Susan Oyama, Michael Ruse, and Kelly Smith. I thoroughly enjoyed our many conversations and benefited enormously from them. Even more in particular: Brian Goodwin visited Amsterdam when I was just starting my project. Our contacts since then have been irregular, but always stimulating and pleasant. I met Susan Oyama at the ISHPSSB meeting of 1991, and I am grateful for her warm interest and thoughtful questions, which have continued ever since.

Many more people have meant a lot to me and this book. Professors Van den Biggelaar and Scharloo, Ger Ernsting, and Peter van Tienderen were interested biologists who gave valuable advice. Then there were the colleagues in the department of theoretical biology, those who are involved in mathematical ways of doing theoretical biology: Cor Zonneveld, Bas Kooijman, Hugo van den Berg, Paul Doucet, Bob Kooi and Jacques Bedaux, who have been helpful in many ways. I am especially grateful to Bas Kooijman for the beautiful pictures he drew for the book. Other people who have helped in various ways are Wim Dictus, Elly van Donselaar, Rolf Hoekstra, Soemini Kasanmoentalib, Theo Kuipers, Thea Laan, Annemarie Mol, Sylvia Nossent, Emilia Persoon, Jeanne Peijnenburg, Hans Radder, Peter Sloep, Bart Voorzanger, Rein Vos, Françoise Wemelsfelder, W. Wilhelm, and Arno Wouters. Finally, there were my close colleagues in Amsterdam. Ad van Dommelen read with great kindness much of what I wrote. Wim van der Steen, who was my thesis supervisor, has been unfailingly trustful, friendly and supportive, as well as quick and dependable in his critical reading.

The land of philosophical wonder about the study of living organisms is a wonderful one, and with the companions I have, it is a joy to travel in it.

1 ✒ Introduction: How to Understand Development?

One way to understand organisms is through their embryonic development. How have they become what they are, starting from the egg? The answer to this question that is common: it's all in the genes. This is not only a layperson's answer, it is also the answer given by many biologists. The embryologist Lewis Wolpert states it clearly: "DNA provides the programme which controls the development of the embryo" (Wolpert 1991, 5). Development is pictured as a hierarchical process of differentiation in which the decisions are made by control genes. Though the picture is not known in all its complicated detail, the overall message could hardly be simpler. As the historian Garland Allen writes, embryology and genetics have been integrated on the basis of the idea that all biological phenomena can be described in terms largely derived from genetics (Allen 1978, 144).

But consider the following phenomena. In 1912, the marine echiurid *Bonellia viridis* was the first organism for which environmental sex determination was described; that is, environmental factors determine whether the animal becomes male or female. Larvae are planktonic and sex develops after settling. If a larva lands on a rock, it develops into a female with a body of about 10 centimeters and a long proboscis, which can become more than a meter in length. This proboscis has a function in feeding, but also in sex determination. If a larva lands on the proboscis of a female, it migrates into the female and develops into a tiny (1- to 3-mm) parasitic male. The male stays inside the female throughout its life and fertilizes her eggs (Bull 1983, 110; Gilbert 1994, 783). See Figure 1.1.

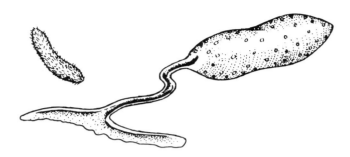

Figure 1.1
Bonellia viridis. Right: female. Left: male, greatly magnified compared with
the female

Environmental sex determination is not the dominant form of
sex determination among animals, but neither is it rare. It takes many
forms: temperature, food, density, or the sex of nearby conspecifics
may be crucial. In many reptiles, for example, temperature is the
factor. This was discovered for a lizard in 1966, and subsequently it
has been found that in many other lizards, in many turtles, too, and
in all crocodiles, the sex of the animal is determined in this way. The
details are species-specific. In some species, females are produced at
high temperature and males when it is colder. In other species it is
the other way around, while females can also result at high and low
temperatures with males in between (Bull and Vogt 1979; Bull 1980;
Bull 1983; Deeming and Ferguson 1988; Lang et al. 1989; Janzen and
Paukstis 1991). See Figure 1.2.

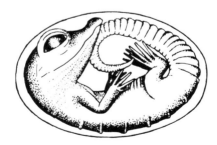

Figure 1.2
A crocodile embryo

Other traits can also be under environmental influence. Many butterflies show seasonal polyphenism; that is, animals which develop in one part of the year, notably the wet season or spring, are phenotypically different from those developing in the dry season or in summer. One of the cases that has been known for a long time concerns the map butterfly or European map, *Araschnia levana*. Its two phenotypes differ so strikingly that Linnaeus classified them as different species (Bink 1992). The spring form is orange, with black spots, while the summer morph is black with a white band. Day length and temperature during the larval period are simultaneously influential (Shapiro 1976). Under experimental conditions with constant temperature, daylength makes the difference. It influences the release of the hormone ecdysone that initiates adult development. Larvae reared under short-day conditions become diapausing pupae and emerge as spring morphs. Long-day larvae become nondiapausing pupae, emerging as the black and white summer morph (Koch and Bückmann 1987).

Other mechanisms are also found. For some butterflies, photoperiod makes the difference, for others, temperature, while sometimes both factors are operative.

DDT and PCBs influence the development of gulls and terns. These chemicals are broken down in the body of the mother into products with estrogen-like effects, which interfere with the normal development of the reproductive organs in male birds. Sterile intersex animals are the result (Fry and Toone 1981; Fox 1994). This example is different from the two earlier ones, in that it involves abnormal development caused by pollutants.

The next and final example is again different, because it does not refer to phenotypic differences resulting from environmental differences, but to the complex specificity of the environment that an animal may need for its development. See Figure 1.3.

Larvae of the insect *Mantispa uhleri* (which resembles a mantid but belongs to the Neuroptera) feed on spider eggs. First instar larvae board immature spiderlings and spend most of their time in the book lungs. They feed on spider hemolymph, and overwinter in the spider. When they happen to find themselves on a male spider, they die, but when their host is female they enter the egg sac when it is produced and resume development there, feeding on the contents of the spider's eggs. The mature third instar spins a cocoon within the spider egg sac,

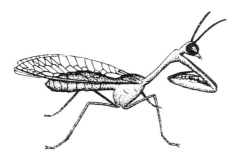

Figure 1.3
Mantispa uhleri

from which the adult animal later emerges. Spider eggs are the obligate larval food for this species (Redborg and Macleod 1983).

Choices

In books that approach developmental biology genetically, for example Lawrence (1992), phenomena such as that discussed earlier get no attention. Gilbert's broader oriented textbook *Developmental Biology* (fourth edition, 1994) does briefly mention various environmental influences. But it is remarkable that they find their place in notes, in special sections called "sidelights and speculations" or, in passing, in discussions that are meant to clarify other aspects of development. Environmental influences evidently have no clear place in developmental biology, and have to be dealt with in the margins.

My aim is to question this situation. First: why are environmental influences almost absent from recent accounts of development, or present only in the margins? Second: could and should something be done about that from a scientific point of view? Third: should something be done about it from a moral point of view?

In the next few paragraphs, I will introduce the various roads traveled in this book to answer these questions. But first, let me sketch the philosophical spirit of the undertaking.

My approach to scientific analysis is a pragmatic one that focuses on the inevitability of choices. I like Stephen Pepper's (1942) characterization of pragmatism, in his book *World Hypotheses*. He distinguishes four "world views," each characterized by a root metaphor. Pragmatism, or contextualism, has the historic event as its root metaphor.[1] It is a "horizontal" world view in that there is no absolute top or bottom of analysis; there are many potentially revealing ways to analyze something. This is its principal difference from the other views, where analysis must be done in prescribed ways. Contextualism rejects analysis for analysis' sake; it is always directly or indirectly practical. As I see it, this holds for scientific, philosophical, as well as any other kind of analysis.

A focus on scientific choices—associated with purposes and with consequences—is the core of my pragmatic approach. Nothing more is involved; I am not defending all the characteristics of the philosophical position termed "pragmatism." In particular, the notion of truth is not central (as it is, for example, in Rorty 1982; see also Murphy 1990); I do not defend a pragmatic theory of truth. Nor do I defend any other theory of truth, for that matter. The concept of truth, in the present book, is taken for granted. Given a vocabulary and given a subject of study, some things are true and other things are not. Scientists try to say things that are true, and this leads to many unproblematically accepted truths as well as many problematic, uncertain, or contested cases. In the context of science, some intersubjective criterion of "that which is agreed upon" suffices to make the notion of truth operational.

Scientists have to choose and define subjects about which they want to find truth, and they have to choose vocabularies and methods. The all-purpose all-encompassing approach does not and cannot exist; the adequacy of a particular approach depends on its purpose. I regard this need for choices as uncontroversial, and therefore the kind of pragmatism that I emphasize is not special or surprising. I just think it deserves more emphasis, in philosophy as well as in science.

Three Approaches to Development

Dressed in this pragmatic "suit," let me sketch how I will address the questions concerning the place of environmental influences in developmental biology.

The first question is: Why are environmental influences almost absent from textbooks on developmental biology? I will answer this question by concentrating on the conceptual situation, not on how this situation has grown. The dominant approach to development is genetic. Within this approach, genes are the primary objects of study, and they are also at the center of the causal picture. Given that picture, attention is not easily drawn to environmental influences; if these are acknowledged at all, they are considered to be relatively unimportant. In Chapter 2, I argue that the mechanisms that exclude the environment are at least partly conceptual.

Though I will not go into the reasons and background of genetic dominance, at least one potentially good reason for it does not hold, which is that a genetic analysis of development might be the only approach that is experimentally feasible. This is not true; other approaches to development are possible, as will be emphasized in later chapters.

The second question is: What could be done about the absence of the environment from explanations of development? A discussion of two different alternatives to the genetic approach provides the background for my answer. The existence of several different alternative approaches is not surprising, for any scientific approach is a complex whole, and there will always be many different potential ways to diverge from it. For the purpose of this book, which is to discuss and highlight the explanatory place of environmental influences in development, I have found it helpful to distinguish three perspectives on development, representing three fundamentally different ways to deal with the environment. The dominant perspective is the genetic one, which is criticized by a structuralist as well as by a constructionist approach. According to both critical positions, the genetic approach concentrates too narrowly on genes. The structuralist approach opposes geneticism by calling attention to the organism as an integrated whole. The constructionist approach takes "organism-environment systems" to be the important units; the boundaries of the organism are not seen as causally fundamental. Thus, the three positions focus on genes, the organism, and organism-environment systems, respectively, as the fundamental units of development and evolution. Main conceptual differences in biology concerning the role of the environment can be elucidated with the help of these three positions. Evidently, such a classification gets frustrated sooner or later. Not all present work in developmental biology fits in precisely with one of the positions. But the positions are not just constructs for the purpose of the discussion; they are defended in the way I present them.

Metaphors and Science

Chapters 2 and 3 are devoted to an exploration of the conceptual landscape through a comparison of the three positions. These are compared in two ways. The first focuses on theoretical choices that are involved, particularly choices concerning causation. Since causal explanation is evidently a core goal of science, the relevance of causal issues does not require much introduction. Chapter 2 discusses causal questions concerning development. Chapter 3 adds a philosophical evaluation. I will stress the unavoidability of theoretical choices and the incompleteness of causal explanations.

The second way to compare the approaches, which is interwoven with the first, is through metaphors. The emphasis on metaphors may require a bit more introduction. Metaphors describe something with the help of something else. Arbitrary comparisons may yield metaphors. For instance, life may be called a "telephone," and this image may generate further thoughts, such as on the place of calls in life, though these thoughts do not necessarily lead to anything fruitful. A metaphor that is actually used for life is *journey*. This metaphor with its associated images such as "standing at a crossroads" is helpful for many purposes; it frames questions in terms of "where to go," for example. The metaphor of life as a programmed machine can do different things, such as stimulating questions on parts that fail when someone is ill. The examples illustrate that a pragmatic evaluation of metaphors, associating them with purposes they do or do not suit, is sensible.

Metaphors are not only important in popularized and immature science, as is sometimes assumed, but in theoretically well-developed parts of science as well. Take the well-developed theory of evolution. Natural selection is and remains a metaphor, however technically the concept may be elaborated. Different metaphorical starting points are feasible, with different technical elaborations. Evelyn Fox Keller rightly notices that "even the most purely technical discourses turn out to depend on metaphor" (Keller 1992, 28). From a pragmatic point of view, the interesting question is if and how different metaphors work out differently.

Metaphors are prominent and irreducible elements in scientific theories; they guide questions, and they guide the integration of data into an overall picture. Donna Haraway's study of organicist developmental biology in the middle part of this century may serve as an illustration. Haraway describes proposals for a nonvitalistic organicism intended to transcend the conceptual devices of the "mechanism-vitalism" controversy. Mechanism saw the organism as a machine. This metaphor was considered completely wrong by Driesch, the vitalist, who thought that embryos are radically indeterminate. The controversy was bitter, but in fact vitalism shared the machine image with mechanism; it only added the assumption that the machine needs a vital substance to make it run. This is the background against which Haraway locates various proposals for a nonvitalistic organicism, which rejected the machine image that had held both sides of the old controversy captivated. The newer approaches centred around different, nonmachine, metaphors such as liquid crystals and fields: "For Harrison the limb field is like a liquid crystal and unlike a jigsaw puzzle. For Needham the embryo is like history interpreted from a Marxist viewpoint and unlike an automobile with gear shifts. For Weiss butterfly behaviour is like a random search and self-correcting device and unlike a deterministic stimulus–response machine" (Haraway 1976, 205). For all these researchers, metaphors embodied basic views of development.

Metaphors color pictures of development from their general features down to details, and pertain to subject definition as well as to views of causation. For example, within a genetic perspective on development, even when environmental influences make a crucial difference, the search is often for the "underlying genetics." "Underlying" and "basic" are persistent metaphorical images indeed in the description of genetics' theoretical place.

Environmental Influence: A General Picture Is Not Enough

A comparison of causal strategies and metaphors clarifies how the three approaches understand development and how they determine what is salient in the process. Both structuralism and constructionism claim to be more complete approaches than the genetic one, suggesting that something is wrong with restrictedness. But a main point of the present book is that none of the approaches is complete. Nor can they be; attention is always selective. Though it is true that the structuralist and constructionist approaches propose to include more factors in developmental biology than a purely genetic one, they, too, through their definitions of the subject of interest and through their causal approaches, depend on restricting choices. Nothing is inherently wrong with this, the important thing is to be aware of the restrictions. Overall frameworks have their limitations, and within these frameworks, the choice of research questions involves further restrictions.

Focused attention is not only unavoidable, it is helpful, even necessary, to get to know a subject in detail. Environmental influences can only have a real place in developmental biology when specific research is devoted to them. In other words, a general picture that acknowledges their in-principle importance in the complex interactive processes is not enough. Let me illustrate this.

In developmental psychology, interaction between internal factors and the environment is widely acknowledged, indeed almost a commonplace. Urie Bronfenbrenner has observed that in spite of this universal approval of interactionism, an internal perspective dominates in psychology; the result is the study of "development-out-of-context" (Bronfenbrenner 1979, 17–21). Bronfenbrenner argues for a broadened perspective that takes into account environmental influences on development. For that end, detailed characterizations are needed of what he calls the "ecological environment," namely, the environment as it is relevant for the developing person—rather than as it may exist in "objective" reality. The ecological environment can be seen as a set of nested structures, beginning with the immediate setting of the developing person, the "micro-environment," up to the "macro-environment," which refers to the cultural or subcultural level. An example of a macroenvironmental characteristic is that crèches look like each other within a culture, but differ considerably between cultures.

The history of developmental biology, too, illustrates that the in-principle recognition of multiple causes does not guarantee that environmental influences receive particular attention. Concerning embryological development, again, nobody will probably deny when explicitly asked that internal factors by themselves are not causally sufficient for the process. There are even those biologists who are very explicit about the insufficiency of internal causes. Oscar Hertwig, who wrote at the end of the nineteenth century, and Gavin de Beer, who wrote in the middle of the twentieth, are among them. Let me quote them, rather than some present-day biologists, in order to give some historical depth to the present discussion. Though both authors assumed that the genetic material contains the potentials of organisms, they warned against a distorted view of causality that sees all other causal factors in development as uninteresting background.

Oscar Hertwig (1894), in his *Zeit- und Streitfragen der Biologie I: Präformation oder Epigenese?* argued, against Weismann, for an epigenetic picture of development. Weismann assumed that each biological character is represented by a determining factor, a part of the "Keimplasma." This way of thinking, Hertwig maintained, rests on wrong ideas about causality. The developing organism is almost pictured as a perpetual motion machine, while in fact numerous conditions must be fulfilled to give the outcome. In normal development, the embryo depends on material exchanges, gravity, light, temperature and so on; these conditions must always be present in the same way but that is no reason to forget about them: "Deshalb dürfen wir aber noch keinesweges die Rolle der Bedingungen, als ob sie gar nicht existierten, ausser Acht lassen, wenn es sich darum handelt, den organischen Entwicklungsprozess ursächlich zu begreifen" (Hertwig 1894, 81).

In a similar way, Gavin de Beer asked in *Embryos and Ancestors*: "Do the internal factors which are present in the fertilized egg suffice to account for the normal development of an animal?", answering that "it may be definitely stated that they are not sufficient, for if a few pinches of a simple salt (magnesium chloride) are added to the water in which a fish (*Fundulus*) is developing, that fish will undergo a modified process of development and have not two eyes, but one" (De Beer 1940, 14).

While De Beer thus stresses that internal factors by themselves are not able to "produce" a normal animal, and are only a "partial cause" of development, the tendency to background the external factors is also clear from his writing: "The internal factors (. . .) enable the animal to react in definite ways to the external factors and by this means give rise to structure after structure in the process of development" (p. 15).

Development cannot take place without an environment: the environment is involved in all traits and in all developmental events. Though this cannot be denied, it recedes easily into the background; a recognition of the general importance of environmental conditions has not resulted in a real place for the environment in developmental biology. To create this real place, study of specific environmental influences is also needed. Chapter 5 contains explicit anwers to the second question of this book—what can be done about the absence of the environment—in the form of examples of environmental influences and ways to study them. One concept that is very useful in the investigation of environmental influence in development is the concept of "reaction norm." It refers to the relation, given a certain genotype, between a range of developmental environments and the resulting phenotypes. Studying reaction norms does not generate complete pictures of developmental processes, but it does yield insights into the environmental dependences of developing organisms.

Hertwig wrote at a time when development was thought to be brought about by either preformation or epigenesis, a distinction deriving from Aristotle. What precisely these terms refer to has been subject to historical change, but by and large preformationist theories hold that development is the unfolding of structures already present in the egg, while epigenesis involves the progressive formation of new structures during development. The present fate of these positions is summarized by Gould, who writes that "modern genetics is about as midway as it could be between the extreme formulations of the eighteenth century" (Gould 1977, 18). Hall, too, writes that neither position has "won out." In one sense, epigenesis has triumphed because embryonic structures are not preformed in the egg. In another sense, preformation is right because the genetic basis for development lies preformed in the DNA of the egg (Hall 1992, 86).

This feeling of synthesis is well expressed by the word "epigenetics." It was coined by Waddington as a better translation of "Entwicklungsmechanik" than "experimental embryology" or "developmental mechanics" (Waddington 1956, 10). In "epigenetics," "epigenesis" and "genetics" are blended. Epigenetics has come to be seen as the causal analysis of development defined as the mechanisms by which genes express their phenotypic effects (Hall 1992, 89).

On the one hand, that "epigenesis" turned into "epigenetics" is illustrative of the central place that genes now have in developmental biology. Though everyone will agree with Hertwig and De Beer that genes are only partial causes of development, they have nevertheless received the far greater part of scientific attention. On the other hand, epigenetics goes beyond pure genetics in that it studies how genes are regulated. The study of gene regulation began with Jacob and Monod's model for the regulation of the lactose-operon in *E. coli*. Jacob and Monod emphasized the relevance of their results for "the fundamental problem of chemical embryology (which) is to understand why tissue cells do not express, all the time, all the potencies inherent in their genome" (Jacob and Monod 1961). Mechanisms of gene regulation show how cytoplasmatic factors are involved in embryology. Thus, Jan Sapp notes that the most important aspect of the operon model for embryologists is "the allowance it made for substances existing in the cytoplasm which are able to switch on or off, or to regulate the action of genes" (Sapp 1991, 246). Therefore, models of gene regulation could silence old controversies over the relative importance in development of nucleus and cytoplasm.

When epigenetics is the study of the regulation of one gene by (products of) another gene it does not widen the scope very much. But the role of environmental factors in the regulation of gene expression also belongs to epigenetics in principle. This is why epigenetics is a field where a genetic and a constructionist approach of development may meet, depending on how epigenetics develops experimentally and conceptually. The issue will return at several points in this book.

Morality

Finally, there is the third question asked at the beginning of this chapter, which is whether something should be done about the absence of the environment in developmental biology from a moral point of view. It is the subject of Chapter 6. Let me sketch the approach.

A pragmatic approach to analysis and explanation could be taken to imply the relativistic view that it does not matter how you approach development, as long as what you do is helpful for whatever you happen to be interested in. But I propose to add a moral dimension by looking at consequences of choices. If science took place in complete isolation from any practical affair, scientific choices and the resulting specific kinds of incompleteness would not have moral consequences. Scientific explanation does have a practical impact, though, in direct and indirect ways. In a world full of problems, perpetual questions arise as to what to do, what to blame, what to change. Causal images are guides for what can and should be done, and the causal pictures that science generates are authoritative guides for what to do, what to blame, and what to change. Alan Garfinkel (1981) has argued these points convincingly. Since scientific views of the facts prestructure moral problems as well as solutions, it matters what causal pictures science generates. So I also agree with Robert Proctor (1991) when he argues for a moral/political philosophy of science, which does not make epistemological but practical questions central. More precisely, in fact, my argument is that these questions cannot be separated.

It matters for practical purposes whether we understand development as a process that is completely governed by genes, or by the whole organism, or as essentially taking place in an environment. Seeing development as an essentially genetic or as an essentially ecological phenomenon gives you quite different starting points when you wonder what it means and takes to live and flourish, and what can possibly go wrong in development. Nothing is wrong with questions about genetic diseases and cures. But those are certainly not the only practical problems concerning development. In nature, things go wrong on a massive scale, by the destruction or pollution of environments. Climate and temperature, too, are relevant. Did the dinosaurs disappear because they had temperature dependent sex determination and temperature changes led to the disappearance of females, for example? And may present-day reptiles be in danger for the same reason?

Relationships with Other Discussions

In this book, approaches to development are mainly discussed and evaluated in terms of the pragmatics and ethics of causal explanation. In the meantime, other discussions are going on that concern development or are relevant to it. Let me sketch relations with some of them.

Within biology, there is much debate on the integration of evolution and development. Embryology was not integrated in the synthetic theory of evolution of the 1930s, and it has long been considered to be evolutionary irrelevant by many. This is now changing rapidly; the view that evolution is constrained by development is gaining influence. Since the disciplines to be integrated are themselves subject to controversy, it comes as no surprise that the character of the integration is controversial. Indeed, the three approaches to development that I distinguish are associated with different views on the integration of evolution and development. In Chapter 4, I will connect this issue with the main focus of this book. A central point is that in neo-Darwinian evolutionary theory, the environment is mainly a cause of selection. This association between environment and selection discourages recognition of a direct causal role for the environment in development.

Another discussion surrounding development, particularly in developmental psychology, is the notorious and never ending nature-nurture discussion. I will hardly ever use the words nature and nurture. Since the issue is so evidently relevant, this silence deserves some comment. Oyama's (1985) book *The Ontogeny of Information* deals with the persistence and many guises of the dichotomy of nature and nurture. The dichotomy has often been declared dead while living on quietly in a different guise. All the guises in some form or other involved the opposition of internal and external causes. A real solution to the controversy, Oyama argues convincingly, requires dropping the internal-external dichotomy. I agree. It is through this internal-external dichotomy, which will show up repeatedly, that my treatment relates to the nature-nurture issue.

A further discussion associated with development is in terms of holism and reductionism. Since the 1920s, when embryology and genetics became separate disciplines, embryology has been associated with holism and genetics with reductionism. What does this distinction involve? In one general sense, reductionism is the claim that wholes can fully be explained in terms of their parts, while holism denies this. This general characterization has different implementations in different contexts. In discussions on development, reductionism often stands for the idea that DNA is or contains a program of development. Embryologists over decades have repeatedly attacked this metaphor, denied the causal sufficiency of the genome, and defended various forms of holism.

In the terms of this discussion, both alternatives to geneticism that I will discuss are holistic: they emphasize the causal insufficiency of the genome. To a certain extent, Chapters 2 and 3 can thus be read as an elucidation of the one reductionistic and two holistic approaches that I have distinguished. The point in these chapters is to explore the approaches and the role they give to the environment. Clearly, the environment can only have a causal role in holistic approaches. But it figures in only one of the holistic perspectives, the constructionist one. Evidently, for the purpose of this book, holism needs further distinction and clarification, and this is one reason to doubt the helpfulness of the term in this context.

Moreover, reductionism and holism have other meanings than the ones just mentioned. In the philosophy of science, for example, from Nagel's analysis (Nagel 1961) onwards, reduction is something that takes place between theories. With respect to biology, the main question has been whether biological theories can be reduced to physical ones.

In other contexts, the terms have been used in still different meanings. Reductionism is often associated with "simplification," for example, and holism with the refusal to simplify, or, less friendly, with vagueness. Debates on the status of reductionistic simplification then may center on the question whether the simplification is only heuristic or has also ontological meaning.

All in all, I have avoided the terminology of reductionism and holism because it causes confusion. Yet some of the problems associated with reductionism—holism debates play a role and I have discussed them, but in different terms. In particular, the heuristic use of reduction (in the sense: reduction of the complexity of a research situation) shows up here as the issue of asking restricted questions, as mentioned earlier. I have been tempted to call such questions "reduced" questions, and to argue in favor of "multiple reductions instead of only genetic reduction," but given the tendency in biology to associate reductionism with geneticism, this might add to the confusion. The term "restricted question" will hopefully avoid such confusion.

Finally, the distinction of overarching approaches to development may invite associations with an analysis in terms of "paradigms," a notion which I will not try to define but which refers to systems of thought that differ deeply and fundamentally from each other. Donna Haraway, in her analysis of the metaphors of organicism in developmental biology, did make use of this term. As she uses it, it refers to situations in which scientists show "partial incomprehension of one another's views and talk past one another on crucial issues" (Haraway 1976, 203). According to this characterization, organicism was not really a paradigm on its own, because much work in the organicist tradition formed a continuity with existing work. Besides, efforts to create a theoretical framework were "often in vague terms that are hard to relate to actual experimental issues" (p. 204), so that this approach was not really full-grown. Nevertheless, the paradigm model helps to see the role of metaphor and imagination in science, according to Haraway.

It is now twenty years later, and metaphors no longer need the help of the paradigm-notion in order to be visible. In the present situation, talking of paradigms might harm rather than help my undertaking, because it is not my intention to stress that the views presented in this book are incomprehensible to each other. There are important differences, but there are also points of overlap, and shared questions, and all kinds of interaction and hybridization may and do occur. This will become apparent in the discussion of the three causal approaches to development, to which I now turn.

2 ✍ Three Causal Approaches

The dominant approach to development is genetic, and it is the first approach I will deal with in this chapter, with a particular focus on the place of environmental influences. Next, I discuss two alternative proposals as to how development should be approached, again focusing on how these proposals deal with the environment. Metaphors and causal questions will be searchlights for understanding the similarities and differences between the views.

Metaphors Surrounding DNA

Let me first introduce the power of metaphors concerning DNA. In the 1950s, David Nanney argued that there are two ways to think about heredity, one in terms of the "master molecule" concept, the other in terms of "steady state." He opposed the dominance of the master molecule concept, or the "Theory of the Gene," which sees the genes as the directors of the activities of the cell and all the other cellular constituents as relatively unimportant except as obedient servants of the masters. The theory of the gene, Nanney wrote, suggests a totalitarian government. The alternative he defended was to see the cell as a more democratic organization "which owes its specific properties not to the characteristics of any one kind of molecule, but to the functional interrelationships of these molecular species" (Nanney 1957, 136; for related metaphors see also Sapp 1987). Nanney's efforts have not been widely successful; if anything the master molecule concept has become more influential. DNA is now almost standardly pictured as the big master molecule that makes all the important decisions in an organism. This image manages to survive many efforts to undermine it.

One such effort is the argument that DNA is not an active molecule; it is regulated by all kinds of factors, not doing very much itself. As Richard Lewontin stated, "DNA is a dead molecule, among the most nonreactive, chemically inert molecules in the living world" (Lewontin 1992a). Remarks such as this are meant to downplay DNA's superpower, and to portray it as a servant that is available to the cell rather than as a ruler (for another example, see Kacser 1960). But the image of DNA as a superpower appears to be compatible with different pictures of exactly how it works; its inactivity need not stand in the way of its power. Consider what Douglas Hofstadter has to say about inactive DNA—not in answer to Lewontin, by the way. Hofstadter is not a biologist but he feels qualified to write on the "logic of the cell," as he calls it. It is a very metaphorical logic. One might regard DNA, he says, as a "big, fat, aristocratic, lazy, cigar-smoking slob of a molecule. It never does anything. It is the ultimate 'lump' of the cell. It merely issues orders, never condescending to do anything itself, quite like a queen bee. How did it get such a cushy position? By ensuring the production of certain enzymes, which do all the dirty work for it" (Hofstadter 1985, 686).

The metaphor here is very explicit and in that respect differs from those current metaphors that are hardly recognized as such any more, such as "genetic program." I mention the example because it uncovers some of the powerful images expressing the central causal place of DNA and genes that underly more covert metaphors. The example also shows that metaphorical thinking is not confined to images about the nature of things, but also involves activities. DNA is not only a cigar-smoking slob of a molecule; it also issues orders, and governs the cell. To put it differently, verbs as well as nouns are in the realm of metaphor.

With regard to development, the metaphor of a genetic program that controls development is prominent. In the first chapter, I introduced Lewis Wolpert who states that "DNA provides the programme that controls the development of the embryo and brings about epigenesis" (Wolpert 1991, 5). This statement is typical of what many biologists now think to be a simple truth about development.

It fits in with the idea that DNA contains all the information and makes all the important decisions in the organism. Development is thought to be essentially a cascade of differentiating decisions, which are made by genes. To mention a few more examples, in *Molecular Biology of the Gene* the question is asked "How, then, can we understand development at the molecular level?" And the answer reads: ". . . the gene products responsible for development can be arranged in a hierarchy, with some genes controlling the expression of other genes" (Watson and Dewey 1987, 748). And the equally authoritative *Molecular Biology of the Cell* states: "The cells of the embryo can be likened to an array of computers operating in parallel and exchanging information with one another. Each cell contains the same genome and therefore the same built-in program, but it can exist in a variety of states; the program directs development along various alternative paths" (Alberts et al. 1989, 901–902).

Terms expressing the power of the genes in development include "master genes" and "master control genes" (Gehring 1987; Day 1990), "smart genes" (Beardsley 1991), and "master chefs" (Schmidt 1994).

Segmentation in Drosophila

The (molecular) genetic approach to development yields many results, and some of them have become paradigmatic examples of the hierarchical genetic control of development. One such case is the development of anterior-posterior segmentation in the fruit fly, *Drosophila*. See Figure 2.0.

Figure 2.0
Drosophila

From the very beginning, a *Drosophila* egg has anterior-posterior polarity: egg, embryo, larva, and adult fly all have a head (anterior) end and a tail (posterior) end.[1] The segmentation story as it is now told in textbooks is a story of hierarchical gene action. First, there is the action of maternal genes. Maternal proteins and mRNA are stocked in various parts of the egg. After fertilization, the proteins— including newly synthesized proteins encoded by the RNAs—diffuse through the embryo, creating various regions and gradients.

Bicoid is prominent among these maternal effect genes. Bicoid mRNA is placed into the embryo by the mother's ovarian cells, and is localized in the anterior part of the oocyte. The bicoid mutant develops no anterior structures, so bicoid protein is said to control anterior development, or, put differently, it is the anterior morphogen, responsible for head and thorax formation. The *Drosophila* embryo contains a bicoid protein gradient with the highest concentrations at the anterior end. This protein binds to and activates the *hunchback* gene, which belongs to the next hierarchical level of gene expression: the gap genes, which are the first genes of the embryo itself to be expressed. Bicoid protein also activates transcription of at least three other gap genes of the head.

Another maternal effect gene is *nanos*. Nanos mRNA is produced in the ovary and placed in the posterior part of the egg. For proper placement, the products of several other genes are needed. Without nanos or any of these other genes, the embryo forms no abdomen. Nanos mRNA is translated into protein soon after fertilization, like the bicoid message. The nanos protein product represses the translation of hunchback mRNA. Thus, bicoid and nanos protein together create a hunchback protein gradient in the embryo. The nanos gene product is said to be the posterior morphogen of the *Drosophila* axis. See Figure 2.1.

Further specification is brought about by more gap genes, and next by pair-rule genes and segment polarity genes. Gilbert (1994) lists the genes known up to then: thirteen maternal effect genes, eight gap genes, nine pair-rule genes, and eight segment polarity genes. The regulation is organized as a cascade; genes higher up in the cascade activate and repress genes lower down. The gap genes are repressed or activated by the maternal effect genes and divide the

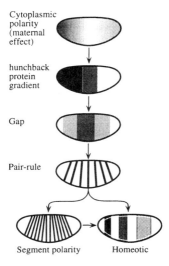

Figure 2.1
Generalized model of the genes involved in *Drosophila* pattern formation (after Gilbert 1994, p. 534; used with permission of Sinauer Publishers)

embryo in broad regions. The pair-rule genes subdivide the broad gap domains into segments. Mutations of pair-rule genes (such as *fushi tarazu*) often delete portions of every other segment. The segment polarity genes give each segment an internal structure. For example, *engrailed* is needed for the anterior-posterior boundary between segments. Finally, the identity of each segment is determined by the homeotic genes, which are regulated by pair-rule and gap genes.

Some fruit flies have legs instead of antennae growing out of the head sockets. Such mutants, with normal body parts arising in the wrong place, were called homeotic mutants by William Bateson around the turn of the century. They are now known to be caused by homeotic selector genes. Two regions of the *Drosophila* genome contain most of the homeotic genes. One is the *Antennapedia* complex, containing five homeotic genes, the other one the *bithorax* complex with three protein coding genes. The position on the chromosome (upstream-downstream) corresponds with the normal place of action (anterior-posterior) of the gene products.

What kind of questions is such a genetic approach answering? Put soberly, genetics does not study the complete process of development, in all its chemical, physical, and other aspects. It looks at the role that genes play, *all other things being equal.* More specifically, genetics analyzes the difference that each gene makes for the outcome, by comparing the effects of different mutations. For instance, to say that bicoid protein is the anterior morphogen is to say that the difference between an embryo with and without a functional bicoid gene is the presence or absence of anterior structures. In other words, bicoid is essential for the formation of anterior structures. But bicoid and the other genes do not work in a void. Many things are implicitly assumed to be "equal." When the story of early *Drosophila* development is presented in purely genetic terms, lots of things could be added to give a more "complete" picture of what is needed, about which genetic research is less likely to ask questions. Take oogenesis, the formation of the egg. The egg is replete with all kinds of mRNA and proteins which are important in the first few hours of the development of the *Drosophila* embryo. It is often said that the maternal genome provides everything necessary for those first few hours. But oogenesis is not completely genetic; the process is influenced by factors such as temperature, nutrition, humidity, and photoperiod (Ransom 1982, 34). For example, yolk synthesis is affected by several interacting agents such as sex, juvenile hormone levels, ecdysone, and nutrition. Juvenile hormone is stimulated by a brain hormone, which in turn responds to photoperiod (Gilbert 1994, 816).

However, in the words and thoughts of many, genetics is not simply looking at the difference that genes make in development, but at the causes of development as such. Genetic terminology suggests that the genes do everything there is to do: they determine, control, specify and so on. Raff and Kaufman, in their *Embryos, Genes and Evolution* (1991) clearly state what kinds of explanations they are looking for: "There is a distinct mental set to our approach to evolution as there is to our approach to development, and this colors our choice of the topics considered in this book. The essential position is that there is a genetic program that governs ontogeny, and that the momentous decisions in development are made by a relatively small number of genes that function as switches between alternate states or pathways" (Raff and Kaufman 1991, xxv). Genes thus function as switches at the decision points, making the difference between one decision and another, and insight into the switch-system is necessary to understand evolution, since evolutionary change, the subject Raff and Kaufman want to understand, consists in changes in this system.

By implication it is also clear what is *not* in need of explanation here: the normal functioning and growth of organisms. The division of a cell into equal daughter cells, for example, needs no explanation by a developmental biologist. According to Raff and Kaufman, the genome can be conceptually divided into two functional parts: genes that are governing development on the one hand, and genes with housekeeping functions on the other (note the social metaphor). The housekeeping functions include everything that is necessary to keep the cells alive and functioning. Mutations affecting these functions would cause severe developmental problems, but would "not necessarily affect a discrete organ, tissue or time of development"; in other words, the effect would be relatively nonspecific. Therefore, they conclude, "such mutants must be distinguished from those of real developmental interest" (Raff and Kaufman 1991, 201; see also Whittle 1983, 63 and Nüsslein Volhard 1994).

So, only genes with specific developmental effects have real developmental interest; the others serve as a background, like all other factors with more general effects. Against such a background of general functions, the decision points in the tree of differentiation are the specific places where (evolutionary) changes can occur.

Mutation is crucial to the work of geneticists, since it reveals the difference that a gene makes. Thus, large-scale study of zebra-fish development started, and had to start, with the production of a large set of mutants. As *Science* reported: "The basic idea is simple: treat adult males with a chemical mutagen and search, three generations later, for embryos that develop abnormally. Because developmental pathways are controlled by cascades of molecular events, however, getting a complete picture of development depends on hitting all or nearly all, of the genes influencing each cascade" (Kahn 1994).

Despite such claims to completeness, the results of genetic analysis are in fact answers to questions such as: What is the (genetic) difference causing the difference between two organisms in one specific developmental character (*all other things being equal*)?

Switches and Responses

The genetic story about early *Drosophila* development has quickly become a paradigmatic example that illustrates how development is controlled. In fact, the story is not genetic but epigenetic: it tells about the regulation of gene expression. In many cases, as in the *Drosophila* story just discussed, epigenetic attention does not extend beyond genes, which is evidently encouraged by the view that genes are controlled by other genes. Sometimes, epigenetics is indeed used as synonymous with "genetic control of development" (Rollo 1994, 61). Indeed, the view that genes control development leads many developmental biologists to disregard environmental influences altogether. This is not necessarily so. But when they are not disregarded, the genetic viewpoint makes them look not-so-important, as we shall see.

C. H. Waddington and I. I. Schmalhausen are often referred to as important forerunners for an epigenetic approach that includes environmental influences. They indeed acknowledged environmental influences in development, and did so in essentially similar ways. Let me deal with Waddington in some more detail. He recognized environmental influences in development, while at the same time he focused primarily on the genome when it came to causal importance.

Genetics and embryology had become separate disciplines in the 1920s under the crucial influence of Morgan. Waddington was one of the first who tried to bring them together again, at a time when many embryologists saw no big role for genes in development. Waddington put much emphasis on the interactive nature of biological processes. Having been influenced and impressed by Whitehead's process metaphysics, he saw process and interaction as central issues in biology. Nevertheless, he writes that in a situation where embryologists employed all kinds of vague notions such as potencies and organizers, he wanted to return to Morgan's idea that the only "potencies" it is meaningful to talk about are the potential activities of genes (Waddington 1975, 9). According to him, "all the properties of the cell are ultimately determined by genes" (Waddington 1947, 52). How can this be understood?

In some recent papers on Waddington's work, Scott Gilbert (1991a, b, c), a prominent developmental biologist himself, has called attention to the way in which Waddington tried to reinfuse genetics into embryology. Waddington and others conducted embryological research in the 1930s. Embryology was under the spell of induction at that time: in 1924, Spemann and Mangold had published results showing that during gastrulation the dorsal lip tissue induces ectoderm to become neuralized. This was a spectacular finding, and embryologists started to look for the nature of the inducing signal, the "inductor" or "evocator." The unexpected discovery that Waddington and others made was that a variety of natural and artificial compounds could induce the formation of neural tissue. Waddington then concluded that the evocator is "merely a differential" and that it is the genetically determined response by the competent cells that is responsible for the details of the developmental process and thus for the kind of tissue produced (Gilbert 1991a; Waddington, 1947, 54). Waddington adds that "Since it is the genes which control the characters of the animal and its tissues, it must in general be the genes which determine the properties of the competence." Therefore, Waddington shifted his attention from the evocator to the response of the "competent cells."

Developmental systems can be thought of as systems of canalized pathways, according to Waddington. Canalization refers to the phenomenon that the wild type is often relatively uniform, which is supposedly due to some kind of correction of internal or external disturbances. The phenomenon is visualized in Waddington's "epigenetic landscape." See Figure 2.2.

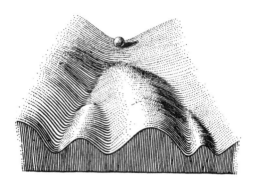

Figure 2.2
Part of an epigenetic landscape (from Waddington 1957, p. 29; used with permission of HarperCollins Publishers)

In this landscape, development of a particular character, or part of the egg, is symbolized by the rolling ball in the figure. Starting in the egg, there are descending and branching developmental pathways that result in distinct phenotypic outcomes, separated by hills where the ball does not roll. That is to say, it can roll there, but there are thresholds to overcome. For example, in the figure, when the ball turns to the right, there is a possibility later on to turn to the left, but this path can only be reached over a threshold. The form of the landscape represents competence.

According to Waddington, the epigenetic landscape, which determines the response to external influences and disturbances, is governed by a large number of interacting genes. The capacity of the organism to respond to an external stimulus by some developmental reaction is thus under genetic control. In the next picture, we look at the space underneath the landscape and see its foundation. The pegs in the ground represent genes. The modeling of the epigenetic landscape is controlled by the pull of the interconnected guyropes, which are the tendencies produced by, or anchored to, the genes. See Figure 2.3.

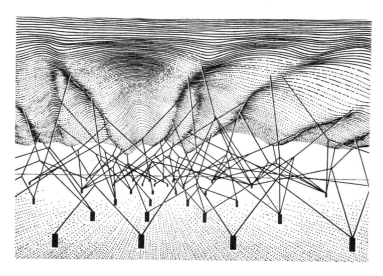

Figure 2.3
Underneath the epigenetic landscape. The pegs in the ground represent genes. They model the epigenetic landscape by pulling the strings, which form a complex pattern and are anchored in the genes (from Waddington 1957, p. 36; used with permission of HarperCollins Publishers)

The environment, according to Waddington (1942), can act in development in one of two ways: either as a switch, evoking an "all or nothing" response (follow a certain path or not), or as a factor that modifies the path, evoking a proportional response. Sometimes it is difficult to decide what should count as a path, but in other cases the alternative paths are clearly defined. For example, in *Bonellia* with its environmental sex determination, the environment acts as a switch between two well-defined alternatives, male and female.

Canalizing processes are processes in which a stimulus that formerly induced a proportional response changes into a switch inducer (by the introduction of a threshold), so that the response becomes "all or nothing." Within a certain range, the response thus becomes unresponsive to the environment. Canalization reduces the causal importance of the environmental stimulus; it becomes a mere switch, and a switch is easily replaced (Waddington 1942, 565). Once a response has become canalized, Waddington says, it is relatively easy to switch development into that track by other factors.

Gilbert adopts Waddington's viewpoint concerning the causal primacy of competence. He stresses that "preset programs," which can be "triggered into action," determine competence (1991c, 354). He even says that competent tissues can "change their triggers," almost suggesting by the use of this active verb that it is the competent tissue itself that determines which factors will trigger it (Gilbert 1991b, 195; 1991c, 351; 1994, 852).

In short, for Gilbert as for Waddington, a switch is merely a switch and its agent can easily be changed, while the specifics of development are in the genetically programmed possibilities of the responding tissue. So the causal logic of switches is that they are not very important because their task is so simple that it can easily be taken over by other factors.

But there are two things to wonder about. First, there is something odd about this logic of switches. Consider what happens, in contemporary biology, when a gene does the switchwork. The switch is then suddenly not a mere switch but a master gene. Gilbert writes, for example, that a particular gene appears to be a "master switch" gene in that it can convert other cell types into muscles (1994, 330). Raff and Kaufman state that the momentous decisions in development are made by genes that act as switches (1991, xxv, see also the preceding section) and Wolpert (1991, 93) says that a gene that switches off some genes and sets in train a sequence of gene activation is appropriately called a master gene. In short, switchers have no consistent causal importance: environmental switchers are *mere* switches, genetic ones are master genes.

Second, if it is true that a switch can easily be replaced, why and in what respect does that make it causally unimportant? The reason cannot be that it can easily be replaced at will; the kind of developmental engineering required would be out of the question in most cases. So the replaceability exists mainly "in principle." Nor can the reason for the causal unimportance of the switcher be that its function can be missed in ontogeny; the presence of the switching factors is vital. If a heat shock produces the same effects as a genetic mutant, nobody concludes that the gene can easily be replaced and is *only* a trigger; instead it is concluded that the effect of the shock copies that of the gene: the outcome is a *phenocopy*. So in what sense is the influence of temperature less important if ether or a gene could do the same work in principle?

Of course, genes and temperature are different. DNA is subject to evolutionary selection in a way that temperature is not. One reason for thinking that genes are in control may be precisely this evolutionary difference. But the fact that the genome is special in evolution does not imply that genes are in control of the process of development. It only implies that together with other factors they give an outcome that is evolutionary successful.

A second background of the assumption of genetic control may be that the environment is often seen as a source of unpredictability, with the implication that reliable development cannot possibly depend on it; it needs an adapted genome. Yet normal development needs a normal environment as well. Genes, together with other factors that are reliably there, enter into a complex process that yields the normal outcome. As we shall see, constructionist authors emphasize that a reliable environment is as essential to development as a reliable genome.

Thus, the effect of Waddington's work is ambiguous. On the one hand, he has called attention to environmental influences in the epigenetics of development, and this aspect of his work is reemphasized in recent arguments in favor of an epigenetic approach to development (Hall 1992; Rollo 1994). On the other hand, through the causal unimportance of triggers or switches, the idea of genetic control remains dominant even in many treatments of environmental influence.

In a more recent example of this perspective, Sandra Smith-Gill (1983) has also distinguished between two different types of environmental influence on the phenotype, which she calls "developmental conversion" and "phenotypic modulation."[2] In the former, development is switched into a particular part of the program, and genes are in control of what happens. In these cases "the organism is genetically adapted to use specific environmental cues to dictate developmental events." In "phenotypic modulation," by contrast, the genome is not in control; the organism is "passively responding to a variable environment." Nonspecific phenotypic variation results from such sensitivity to the environment because of "lack of mechanisms to completely block out or regulate external stimuli" (Smith-Gill 1983, 50). While conversion involves discrete switches, modulation is characterized by a gradual response that ranges from minor effects to death-causing perturbations or the formation of monsters.

In response, first, the adaptiveness of conversion-type responses cannot be taken for granted a priori. In general, assumptions of adaptedness require an insight in the consequences of the phenotypic outcome (Sultan 1987). Second, the example again illustrates how even in treatments of environmental influence causal emphasis may be on the genome, for instance when it is said that conversion is under genetic control and that the genome *uses* environmental inputs. Many treatments of phenotypic plasticity, which is the general term for variation of the phenotype under environmental influence, show this emphasis and are especially looking for the "genetic basis" of plasticity. The assumption is that lack of genetic control is dangerous.

According to Waddington, parts of an embryo are always in dialogue. But his causal treatment of switches encourages views in which the dialogue is hardly real. Wolpert for instance, using the conversation metaphor when he talks about signals between cells, remarks that the conversation is in fact uninteresting. The reason is precisely that inducing signals merely select possibilities from the cell's internal genetic program (Wolpert 1991, 38–47; Wolpert 1994).

So much for a genetic approach to development and its consequences for the treatment of environmental influences. I will discuss two alternative approaches that are critical of the dominance of genetics. The alternatives have things in common; also, they both have points of similarity with the genetic approach, so neither of them is completely or absolutely different from the genetic perspective. But the three positions all differ from each other with regard to their treatment of the environment.[3]

Fields /Structures: Goodwin versus Neo-Darwinism

According to Brian Goodwin, the great challenge for developmental biology is to explain biological form. This makes his approach different from a genetic one right from the beginning, since genetics is primarily oriented toward (biochemical) tissue differentiation. The emphasis on form fits in with a tradition in which D'Arcy Thompson's *On Growth and Form* of 1917 is a prominent landmark and which aims to explain form in a holistic and precise way.[4] The approach makes use of mathematical devices that can explain the formation of spatial patterns, such as reaction-diffusion equations. The primary intention is often to explore formal possibilities rather than to identify the material realization. Some would hold that this approach need not conflict with a genetic developmental biology and that an integrative view may emerge in the future when mathematical variables and gene products can be brought together.[5] Others claim that the approaches rest on conflicting causal foundations. Goodwin, on whose work I concentrate, is outspoken in this respect. According to him, the approaches are fundamentally different.

Goodwin's opposition to the dominance of genetics in developmental biology goes back at least as far as 1970, when he wrote that the genetic program metaphor has unfortunate consequences. This metaphor suggests that genes and the "information" they contain constitute sufficient explanations of development. This makes biology look different from physics, as if in biology only information processing matters, while in fact biology, too, should be concerned with the organization of matter and with fields of force (Goodwin 1987, 335).

In contrast to most geneticists, Goodwin explicitly raises the issue of how to deal with causation in biology, arguing that the neo-Darwinian style of causal explanation is mistaken: it is historical rather than scientific. Ever since Darwin, biology has been an essentially historical science that explains contingent differences between organisms. Consequently, it has not enough to say about the mechanisms that determine how organisms are generated. A scientific approach looks for such mechanisms, which are governed by atemporal laws of nature; indeed, Goodwin consistently downplays the time factor in biological explanation.

In biology, genetic explanation is the dominant form of historical explanation. Genes are historically contingent causes; they are transmitted from parents to offspring, and mutations change them in the course of evolution. Such historical causes, says Goodwin, are necessary but not sufficient. As genes can yield only very incomplete explanations, the idea of genetic programs exercising ultimate control should be rejected. As an example, let us see how Goodwin criticizes a genetic explanation of the presence of six toes on the limb of a cat. The element of truth in the genetic explanation, he says, is that the *difference* between a normal five-toed cat and one with six toes correlates with a genetic difference. However, though that difference may loosely be said to cause a difference of form, this is not the same as explaining how the form is generated (Goodwin 1984a, 106).

Goodwin is right when he observes that since Darwin, many (evolutionary) biologists are mainly interested in historical explanations of differences. Darwin himself set the trend, explicitly saying, in his chapter on instincts, that "I have nothing to do with the origin of primary mental powers, any more than I have with that of life itself. We are concerned only with the diversities (. . .) of animals within the same class" (Darwin 1859/1964, 207). Indeed, in the view of many evolutionary biologists, it is diversity that should be explained. Stephen Gould writes that because of this, evolutionary biology is not looking for laws of nature, but for endless detail. Laws of nature may perhaps set the channels within which organic design must evolve, but the channels are simply too broad "relative to the details that fascinate us" (Gould 1991, 289). At this level of detail, he says, contingency dominates, and the impact of laws recedes into an irrelevant background. The question of why humans exist, for example, has as an important part of its answer the historically contingent fact that the chordate *Pikaia* survived the Burgess decimation (the extinction of most Cambrian groups), and, Gould adds, no "higher" answer can be given (Gould 1991, 323).

From Goodwin's perspective, this way of looking is very incomplete and limited. If you want to explain evolutionary change, the emphasis on explaining differences may be enough. But if you want a complete account of the order and the patterns underlying diversity, the emphasis on diversity takes you nowhere. From that point of view, genetics-dominated biology is too little of a natural science. The alternative is that biology must explain the origins of form in organisms. Such explanations are more complete as they focus on the principles that govern developmental processes.

The causes of development, according to Goodwin, do not reside in the materials that make up the organism, but in the structure of the organism. Organisms are wholes which are best understood by seeing them as morphogenetic fields.[6] A field is a spatial domain in which every part has a state determined by the state of neighboring parts so that the whole has a specific relational structure. Disturbances to the field are followed by restoration of the normal order, by regulative behavior of the field. Fields, not cells, are the basic biological entities, and the most important biological laws are those governing transformations of fields. Thus, explaining morphogenesis is explaining systems of transformations of fields. The organizational principles that govern these transformations should take the form of (mathematical) field descriptions. Genes play a role by determining the material composition of the field, thereby setting certain parameter values for the field equations (Goodwin 1984b, 1985, 1988a, b, 1990a). The role of the environment is the same: setting the values of parameters. While the values of parameters must be within certain boundaries, they do not determine the nature of organisms because this nature lies not hidden in the DNA or elsewhere but is contained in the relational structure of the organism.

Goodwin's favorite example concerns the unicellular green alga *Acetabularia* (Goodwin 1990b, 1994a; Goodwin et al. 1993). He explains the formation of rings of hairlike structures (whorls) in *Acetabularia* not by reference to genes, but by equations with variables representing mechanical properties of the cytoskeleton, the elasticity of the cell wall, and particularly the calcium concentration in the organism. The equations have about twenty further parameters.[7] When computer simulations were done on the basis of these equations, with the values of the parameters within certain boundaries, patterns

were formed that look much like the structures seen in the living
organism (see Figure 2.4, right side).

 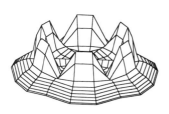

Figure 2.4
Acetabularia acetabulum. Left: a mature cell with a cap and a young cell
with three whorls. The whorls drop off when the cap has developed
(drawing Bas Kooijman). Right: outcome of computer simulation: a ring
of calcium peaks, reminiscent of the ring of hair primordia in a whorl
(from Goodwin 1990b; courtesy of Goodwin)

The field is causally primary, the values of the variables and
parameters that select the solutions of the field equations are secondary. Therefore, genetic and environmental factors do not primarily
cause the organism, because they do not define fields but only realize
or stabilize some of the possibilities of the fields. The real causes of
development are "the laws of physics and chemistry as expressed in
processes that are characteristic of living organisms" (Goodwin 1988b,
635).

What does this imply for the conceptual place of environmental
influences, such as environmental sex determination? Environmental
sex determination can be accommodated easily along these lines: it is
a case where a relevant parameter value in the field equations is set by
the environment. The real challenge, here and elsewhere, remains to
find the field equations (see Goodwin 1994a, 38-39).

Very little is known at present, Goodwin admits, about the set of laws and the integrated theory that are the ultimate goal of developmental biology. But as they become known, it will become clear which organismal forms could potentially be realized, and all the various organisms will have their distinct place in morphospace, the classification of possible forms.[8] If evolution is pictured as a journey, it is thus not a contingent journey through a vast *terra incognita,* as in neo-Darwinism, but a constrained one through the set of possible forms. Evolution is then the time-dependent exploration of this set of time-independent possibilities under internal (genetic) and external (environmental) parametric variation (Goodwin 1989a, 96; Goodwin 1994a, 142–43).

By focusing on the generative principles of organisms, developmental biology should restore the organism to a central place in biology. Goodwin stresses the autonomy of organisms, and points out that his view is similar to the autopoiesis approach of Maturana and Varela (Goodwin 1994a, 162). According to those authors, autonomy or organizational closure is distinctive of the living state. The environment cannot specify changes but can only trigger them (Maturana and Varela 1987, 131). Clearly, such triggering is causally just as "merely" as in genetic approaches.

Summing up, the questions Goodwin wants to answer concern the laws that govern the transformation of morphogenetic fields and define the set of possible forms in both ontogeny and phylogeny. The emphasis in this holistic approach is on physical and chemical determinants of developmental patterns.

Networks/Constructions

A second kind of opposition to genetic programs is exemplified by the views of Susan Oyama, and to a great extent also Richard Lewontin.[9] I call their approach constructionist, after one of their central metaphors, which is meant to express that organisms are constructed in contingent historical processes. This emphasis on time and contingency is a central difference with Goodwin's stress on timeless lawfulness.

Oyama's approach is conceptual. In her book _The Ontogeny of Information_ (Oyama 1985, see also Oyama 1989), she deals explicitly with the notions of (genetic) program and control. The notion of program, she says, wrongly suggests that there is something behind the process that directs it; it is an unwarranted duplication of the order of the process itself.[10] Control in developmental processes is not something behind the process, but is rather in the continually changing influence of heterogeneous interacting elements, environmental as well as internal.

Oyama, like Goodwin, is convinced that with regard to development new central concepts are needed as alternatives to notions such as "genetic program." She favors metaphors such as "construction" and "network." Apart from suggesting contingency, these metaphors are meant to emphasize that causal power cannot be attributed to any kind of (internal or external) factor, but should be associated with interaction. In this view, the very distinction between intrinsic and extrinsic control or causation is misguided. Development must be understood "not as internal to the organism, and certainly not as some cover term for genetic programs, but rather as organism-environment complexes that change over both ontogenetic and phylogenetic time" (Oyama 1992b, 226). Therefore, the central unit of development and evolution is not the organism. It is a larger system that includes parts of the environment. Evolution is change in the constitution and distribution of such developmental systems.

Others similarly emphasize such "decentralization" of control and the causal irrelevance of the boundaries of organisms. In _Not in Our Genes_, Rose, Lewontin and Kamin (1984) criticize the juxtaposition of organism and environment, which, according to them, involves an alienation of the organism from its environment. This alienation is not only a characteristic of theories that see development as an unfolding determined by internal factors. It is also persistent in those "interactionist" views which see evolution and development as a process of trial and error, or challenge and response, where the organism generates "conjectures" in the form of pathways, and the environment selects from them. "Interactionism is the beginning of wisdom", they write (p. 268), but they see such interactionism as flawed nevertheless because, like the older view, it wrongly portrays both organism and environment as autonomous and static. According to them, organisms and environments "interpenetrate" (Rose et al. 1984, 272). Organisms change their environment, environments change organisms. They use the metaphor of dialogue to characterize the relationship. While these authors thus reject interactionism as they associate it with static entities, the term "interaction" is nevertheless often used in constructionist contexts; it does not bear static connotations for everyone.

In a constructionist perspective, the organism is not the central causal unit. In his foreword to the book *Organism and the Origins of Self,* Lewontin (1991) argues against a heavy emphasis on individual organisms as autonomous entities, which he sees as a symptom of the ideology of individualism. This ideology sees organisms as a result of internal forces, which are in competition with respect to the environment: the environment selects among them. In contrast to Goodwin, Lewontin is thus not arguing for a new emphasis on the organism as a remedy against reductionism: "When those who react against the utter reductionism of molecular biology call for a return to consideration of the "whole organism," they forget that the whole organism was the first step in the victory of reductionism over a completely holistic view of nature" (Lewontin 1991, xvi). Organism and environment are co-constructing each other. From a causal point of view, there is no opposition between internal and external forces.

Yet the picture concerning the place of organisms in biology is not so univocal as it seems from these statements. For example, Lewontin stresses that the organism is the subject and object of evolution, and he objects to the opposition of internal and external forces because it makes the organism paradoxically disappear as an active agent. Johnston and Gottlieb (1990), who also approach development in a constructionist way, even emphasize that their theory focuses on the organism. Oyama, too, has insisted on doing justice to the organism as an active agent, deploring a biology in which the individual looks like an abstract nexus of inner and outer causation. To see the organism as a locus of agency is not to deny causality or the social integration of organisms, she writes (Oyama 1988, 270). As she sees it, it depends on the context of discussion or study where the emphasis should be.

So a confusing picture emerges as to the place of the organism. The confusion arises, I think, because organisms have different functions to fulfill in these writings: they may be units of activity, or of change, or of causation. In some respects indeed, it is consistent with a constructionist approach that organisms are central: phenotypes (organisms) instead of gene frequencies are the units of *activity* that the approach stresses. As units of activity they can change in nongenetic yet evolutionary significant ways, for example by taking on new habits. On the other hand, from a *causal* point of view organisms are not the relevant units: causes of development reside not only in the organism, but also in their environment.[11] When it comes to the causal part of the proposals, then, Johnston and Gottlieb's emphasis on the organismic character of the approach is misleading; a focus on inner-organismic causes of development is precisely what they oppose.

Important metaphors in this approach are construction, network, and again: dialogue. Not law-governed fields, as with Goodwin, but historically contingent interaction and temporal sequence are emphasized. In this respect, the constructionist approach resembles neo-Darwinism and genetics. But what is wrong with genetics, according to constructionists, is that it selects only one type of causal elements from processes that are in fact controlled by heterogeneous factors. In terms of the dialogue metaphor: a geneticist approach listens to the voice of only one type of speaker.

Besides, though contingency is a prominent notion in both the neo-Darwinian and the constructionist view, Oyama has pointed out that the views nevertheless differ on this point because neo-Darwinism sees contingency only in evolution, not in development (Oyama 1995). She has argued this point in response to Gould's treatment of the subject. Gould's (1991) discussion of *Pikaia*'s survival of the Cambrium as the cause of chordate existence is the point of departure. The existence of the phylum Chordata is an evolutionary accident, according to Gould, and contingency is the watchword in evolution. Evolution is contingent because numerous different things may happen, which cannot be predicted beforehand, only explained afterward by historical narrative. So he equates contingency with unpredictability. The implication of this view seems to be that what can be predicted is not contingent. Indeed, the usual assumption is that development, which is a predictable process when it proceeds in its normal way, is not contingent. Centrally controlled causal processes are then often assumed to be necessary in order to account for the reliability of the process.

But, says Oyama, while predictability is an epistemological notion, contingency is ontological, it concerns the nature of processes. These different notions need not coincide; in fact, it is better to distinguish clearly between them. Highly predictable outcomes of development can be contingent in the sense that they are not absolutely necessary, or that they are dependent on factors that may be uncertain. In short, contingency can have different meanings, and unpredictability is only one of them.

When the epistemological issue of predictability is distinguished from ontological issues regarding contingency and necessity, the predictability of development need not be a reason to postulate special causal devices generating necessity, such as programs or plans or any other controlling agency. If various kinds of developmental resources of a contingent kind are available, none of them in control of the process, the outcome can still be predictable; this is the case when all the normal resources are reliably present. Environmental circumstances, for instance, which are often regarded as accidental and unpredictable, are in fact highly reliable in most cases. Each generation anew depends on interactions between insides and outsides, and each time all interactants must be there for the normal outcome.

So, while Goodwin deplores the emphasis on contingency in neo-Darwinian explanations, Oyama thinks that we need more of it: neo-Darwinism highlights contingency in evolution but neglects the contingency of development.

It is not hard to see how environmental sex determination is approached within this perspective. It represents a nice example of distributed control; internal and external factors cooperate to bring about the end result.

More generally, what is the task for developmental biology? What kind of explanatory questions must be answered? In this perspective, there is no definite answer to this question. The emphasis in the work of Oyama and various other authors in this tradition is not on explanation of specific phenomena, or a specific explanatory program, but on clearing the road of many old conceptual obstacles. The aim is to make room for a more interactionist picture of evolution and development and a systemic account of causation that can guide research questions.

Many specific questions as they are posed now are too one-sided, according to constructionist authors. Both narrow geneticism and narrow environmentalism must be rejected. A "reversed picture" of development in which environmental causes are privileged is theoretically just as backward and simplistic as the genetic picture, as it is grounded in the same internal-external dichotomy, which makes the organism the passive consequence of competing determinisms. Instead, causal attention should be given to interactions, and system-boundaries should be seen as relative. Since the effect of any internal or external influence depends on the larger system, a developmental system includes all the nested contexts in which a piece of DNA has its effects. Boundaries can be drawn for the sake of analysis, says Oyama, but they are always provisional, and there are always other ways to draw the lines. Likewise for causes and effects; depending on the purpose, causal status can be given to particular aspects of the process. The system as a whole, though, does not consist of discrete causes and effects, but of ongoing processes (Oyama 1994).

Russell Gray also opposes one-sided stories. He discusses Bateson's response to Dawkins' view that an animal is just a gene's way of making more genes; well, a bird is just a nest's way of making more nests, Bateson responded. According to Gray, reversing gene privileging by talking about an "environment for a trait" instead of a "gene for a trait" is surely a nice way of showing the arbitrariness of singling out any factor. But however amusing such environmentalist inversions might be, inverted narratives that privilege the environment rather than the genes are not to be applauded.[12] A constructionist perspective equally opposes environment-centered stories and gene centered stories, because it is relationships, not entities that count: "The basic error of these entity based accounts is that they take networks of co-defining, co-constructing causes and attribute control to just one element in the network" (Gray 1992, 194).

In sum, this direction of constructionist thinking stresses a general approach to causal analysis rather than specific questions. As a general view it seems to me very right. However, I think that the great reluctance of constructionists ever to ask one-sided questions may invite unnecessary paralysis, and may prevent asking much needed questions about the specific influence of environmental circumstances. I will return to this point in Chapter 5.

Comparison and Conclusions

Both morphogenetic fields and constructional networks are proposed as alternatives for genetic programs. Yet there are similarities as well as differences among the three views discussed (see Figure 2.5). Pairwise similarities are schematically summarized in the scheme below.

1. Inside-outside. Neo-Darwinism fits in a long tradition that understands development as coming from within. Goodwin's structuralism does not address this issue explicitly, but implicitly joins the internalism of the tradition. Constructionists want to break with this tradition. They see organism-environment systems as the units of evolution and development; the boundaries of the organism are not causally fundamental.

2. Contingency. Constructionists also have something in common with geneticists: their causes are contingent in principle; no general laws are involved in their explanations. The

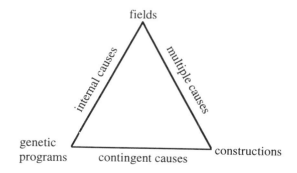

Figure 2.5
Similarities and differences between the three approaches

metaphor of conversation used in these perspectives likewise suggests (by association rather than by logical force) the importance of time, history, and contingency. For Goodwin, historical contingency is secondary; ahistorical necessity defines the scientific content of biology.

3. Relational causation. Both structuralists and constructionists stress the relational and systemic nature of causation. Genes are not enough, but moreover, the development of an organism is a dynamic process in which no discrete causes and effects can in fact be distinguished. They proceed to elaborate this holistic main idea in different ways: structuralists are interested in field explanations with a large emphasis on physical forces. Constructionists emphasize that the boundary of the organism is not causally fundamental.

Each approach presents a different causal picture of development. How do these pictures relate to each other? Can we look forward to a future victory of one, or an integration of all? Kelly Smith, dealing with two of the three views, suggested the second possibility in a paper on *Neo-Darwinism versus structuralism* in which he aimed "to point out what each school can learn from the other and in doing so, help set the stage for a more complete theory of biological order" (Smith 1992b, 434).

While Smith suggests that completeness can be found by some kind of integration, the approaches have a tendency to suggest that theirs is already the most complete approach in principle, and that an integration of data provided by other approaches is unproblematic, again in principle. Genetics can "integrate" all kinds of causal influences by talking about constraints and background conditions, structuralism can "integrate" genetics and environment by talking about parameter settings, and constructionism can "integrate" genes as one type of interactants (on laws they are rather silent). So each approach foresees completeness and integration on its own terms.

Completeness is an important ideal in science, and the field of embryology is clearly no exception. Approaches claim completeness, and when they are found wanting in this respect, an overarching integration is proposed as a remedy. In the next chapter, I will again deal with the relations between the approaches, focusing on philosophical issues surrounding completeness and on the heuristic role of theoretical perspectives.

3 ✑ Explanations in their Theoretical Context

Scientific explanations are quite diverse; in biology, the existence of functional explanations alongside causal ones is an illustration. I will limit the discussion to causal explanations, but within that category, too, much heterogeneity exists. In the philosophy of science, starting with Hempel and Oppenheim's notion that to explain is to subsume under a general law, there have been many attempts to cover explanation with a general model. The undertaking is notoriously unsuccessful. As a result, there are not only many types of explanation, but also many general models that explicate what an explanation is, or really is, or should be. I will not endorse any such model or try to add a new one; different models are illuminating for different purposes or in different situations (see also Kuipers 1986; Van der Steen 1995). Approaching both biology and philosophy in a pragmatic way, I take it that different kinds of biological explanation represent different kinds of knowledge, with a usefulness depending on their purpose and their consequences. Similarly, philosophical models of explanation illuminate explanation in different ways, their potential usefulness again depending on the context.

A pragmatic view of philosophical tools implies that there is no fixed approach to science, not even to the pragmatic aspects of science. The usual pragmatic approach sees explanations as answers to questions and assumes that the specific, often hidden, character of questions is context-dependent. This analysis is helpful in many cases, but a pragmatic philosophy should also look at its own analytical devices in a pragmatic way: if there is no absolute way of analyzing anything, this applies to the understanding of explanation as well as to explanation itself. In philosophy, too, there are always choices to be made.

Analysis of causal explanation in this chapter is not a goal in itself but a means to understand perspectives on environmental influence in development, as discussed in Chapter 2. How can environmental influences be valued so differently? Would it be possible to elaborate a complete and overarching perspective in which every conceivable aspect of development finds a place, or is it perhaps more advisable to choose one of the perspectives? There is broad consensus that in practice, causal explanations are never complete; they highlight some factors as salient, and are silent on many other things. But maybe some complete explanation exists in the background?

I will first discuss philosophical explications of complete explanation, and argue that explanations should be understood in their theoretical context. Theories do not represent complete perspectives. Therefore, a more natural question becomes how theoretical perspectives can be compared and evaluated. How do they differ, and how do they create salience? I will argue that peculiarities and limitations of particular perspectives on development are not in themselves a problem as long as they are acknowledged.

Completeness and the Ideal Explanatory Text

Complete explanation is a scientific ideal that keeps philosophers of science wondering. In Hempel and Oppenheim's account of explanation (1948), further defended and developed by Hempel (1962, 1965), an explanation is an argument. The explanandum (the item to be explained) is the conclusion of the argument, while the premises comprise the explanans; they consist of one or more general laws and the initial conditions. Hempel distinguished between deductive-nomological explanation, in which the conclusion is inferred with certainty, and probabilistic explanation, which makes the conclusion probable in some degree. Let me restrict the discussion to the first kind of explanation. A full or complete explanation includes all the relevant facts and all the relevant laws, and also reveals the logical argumentative steps that lead to the conclusion, such that the explanans implies the explanandum. In practice, explanations will never be complete. The relevant laws, for example, are often omitted in the explanation. Such explanations are elliptical, says Hempel: the laws are tacitly there, and they could be added in principle, to make the explanation complete. Further, explanations will often be partial, that is, the explanans does not completely explain (that is, logically entail) the explanandum as it stands, but explains some wider group of phenomena.

Hempel stresses that the completeness of explanations is relative to the explanandum sentence. An explanation cannot be about concrete events, since any event has "infinitely many different aspects or characteristics, which cannot all be accounted for by a finite set, however large, of explanatory statements" (Hempel 1962; also in Ruben 1993, 27). Hence, explanation is not about the world but about descriptions of the world. Language and the state of knowledge are the relevant domains.[1]

Among the discussions and criticisms of this approach, Wesley Salmon's contribution has been influential. Salmon distinguishes between epistemic, ontic and modal approaches to explanation. (I will disregard modal approaches.) Hempel and Oppenheim's approach is epistemic: it concerns human knowledge and reasoning. Salmon favors an ontic approach, in which explanation refers to events and to the causal structure of the world, not just to our theorizing and reasoning about them. According to Salmon, explanations cannot be mere arguments because arguments do not always refer to causes. Indeed, the requirement that laws are general does not imply that they involve causes of the explanandum; they may just be empirical generalizations.

Since Salmon's work probably forms the most serious recent effort to spell out the idea of complete explanation, let us look at it in some more detail. In Salmon's causal ontological approach to explanation, completeness does not refer to deduction from general laws, but to knowledge of causal mechanisms. For any event, the "ideal explanatory text"—a notion that Salmon takes from Railton— would spell out the complete set of such mechanisms. Since the ideal text would contain all the relevant causal and lawful connections (including "molecules, atoms, subatomic particles and interactions") involved in the event under explanation, it will always be "brutally large and complicated," and it is easy to see that it may never be realized, Salmon says. But, he adds, that does not matter. Though the ideal explanatory text may never be realized, it constitutes a framework that provides guidance to scientists who want to fill out parts of it, and that is its primary function (Salmon 1989, 159–160).

While Salmon is extremely interested in explanation, he stresses that *understanding* on the other hand is not his business. "Understanding" is an extremely vague concept that can easily lead to "anthropomorphism" in the philosophy of explanation (p. 127), which, apparently, is to be avoided in his ontic approach.

The Ideal Text and Pragmatism

Salmon finds the distinction between ideal explanatory text and explanatory information fruitful in that it can help to reconcile the views of pragmatists and realists. Realists are interested in the ideal explanatory text, while pragmatists select salient parts of it as "explanatory information." While the ideal explanatory text determines in a nonpragmatic way what is objectively relevant for an explanation, in the selection of salient parts "pragmatic considerations immediately loom large" (Salmon 1989, 161).

This embedding of pragmatics in the ideal text is important for Salmon because it guarantees a foundation of objectivity. If pragmatic considerations were ultimate, anything would go and there would be no demarcation between scientific and other explanations. For that reason, pragmatic information must coincide with parts of the ideal text, which determines, as a matter of objective fact, what is relevant and what is not. For example, an astrological explanation of President Kennedy's death can be ruled out, since such an explanation "fails to coincide with any part of the ideal explanatory text," he writes; "The configuration of stars and planets and satellites at the time of Kennedy's birth is (I firmly believe) irrelevant to the date of his assassination" (Salmon 1989, 162). The ideal text thus affords "a yardstick against which to measure the explanatoriness of proffered explanation" (Railton quoted by Salmon 1989, 160).

But how can the ideal explanatory text be such a firm criterion for distinguishing between relevant and irrelevant scientific information, when it is not known? The answer is, I think, that it cannot. Salmon's firm belief that astrology is not a canonical field of causes cannot be based on an unknown ideal text, it can only be based on the scientific theories that he endorses.

Now for Salmon, theory is on the same ideal level as ideal texts. The ideal theory itself is not affected by pragmatic considerations; pragmatic considerations only "pick out" things from the objective causal net. He can present this picture as a pacification between ontological and pragmatic approaches to explanation because his arch-pragmatist, Van Fraassen, appears to share its presuppositions, if not in their ontological form then at least in their completeness-assuming form. For Van Fraassen, too, explanations take place against an idealized theoretical background. Ideal theory for him is empirically adequate instead of realistic, but that does not change its ideal character (Radder 1989).[2] In his attack on the realistic character of theories, he says that the interpretation of theories does not require pragmatics. Pragmatic factors, which relate to the speaker or the audience, only play a role in theory appraisal, and in the explanatory use of theories (Van Fraassen 1980, 91). Pragmatics, for Van Fraassen, refers to persons and their specific situations. "Theory appraisal" may be involved, but not theory itself.

However, what if that ideal level beyond theory appraisal does not exist? The idea of an ideally complete scientific theory, something beyond the level of human understanding and error, is a far cry from theories as they are found in scientific practice. Theory appraisal, including the framing and evaluation of theories, requires all kinds of choices and those choices are inevitably human; theories are not ideal texts but human constructs. Therefore, pragmatic considerations concerning "what to pick out" apply to theories, not only to salience within theories.

Contrary to Salmon's fears, however, this does not imply arbitrariness. The existing body of accepted theories and the mechanisms of criticism within scientific communities ensure that not anything goes. Science is constrained in many ways, for example by criteria of truth. Of course, these mechanisms are not offering any absolute ontological guarantee, since they are mechanisms of human judgment. But neither do they turn explanation into something arbitrary; theories are not arbitrary whims.

Let us turn to the pragmatics of theories.

Trade-Offs

There are many requirements for an ideal theory of the Railton-Salmon type: it should be precise, realistic, coherent, general, well-confirmed, comprehensive, and so forth. But no theory can fulfill all such requirements, as Richard Levins (1966, 1968) has argued: "There is no single, best all-purpose model. In particular, it is not possible to maximize simultaneously generality, realism and precision" (Levins 1968, 7). Writing about models in applied ecology, he noticed that such models usually sacrificed generality to realism and precision, while physics-oriented biologists tended to sacrifice realism to generality and precision. In response to criticism (Orzack and Sober 1993) he later explained that he never meant those criteria to be formal or exhaustive. He had been thinking of the special decisions that population biologists were facing at the time (the middle of the 1960s), but not with the intention to present them as the only possible decisions; "what was important was the notion of trade-offs in model building" (Levins 1993, 547).

Levins is primarily concerned with mathematical models. But these insights have a much wider relevance. Wim van der Steen has argued that the idea of trade-offs is a key insight that should apply to all science and philosophy. No theory can satisfy all methodological criteria; scientists are continually forced to face trade-offs. It depends on the purpose of the model or theory how such trade-offs can best be made. That is, pragmatic considerations are fundamentally involved in the framing of theories (Van der Steen 1993a, 103; 1993b).

What can be said of the three causal approaches to development with the idea of theoretical choices and methodological trade-offs in mind? Trade-offs are clearly there. From a methodological viewpoint, different scientific virtues are maximized. The genetic approach tries to maximize realism at the molecular level, downplaying the need for general laws. Structuralism aims to maximize lawful generality, downplaying the need for molecular detail; constructionism can probably be said to aim at realism on the level of interactive causation, sacrificing biological generality.

Linguistic Choices: Metaphors

Descriptions of events are linguistic entities, as Hempel's account of explanation does, but Salmon's doesn't, acknowledge. James Fetzer has brought up this point in a review of Salmon's (1984) book *Scientific Explanation and the Causal Structure of the World*: "As Hempel has remarked, events are never subject to explanation as 'bare particulars' but only as events of a specified kind (under a certain description), thereby presupposing linguistic relativity: without language-dependence, there would be no 'explananda' to explain!" (Fetzer 1987, 600). Fetzer stresses that the linguistic relativity of scientific inquiries cannot be circumvented, because the world does not come to us in a "prepackaged" way. I consider this argument to be entirely convincing, in fact almost self-evident in contemporary philosophy. Since there are always different ways to describe an event, the important and immediate consequence is that no single ideally complete explanatory text for an event exists or can exist.

Though this unavoidable linguistic relativity does not depend on the presence of metaphors, metaphors illustrate it vividly. It is increasingly acknowledged that the scientific perception of similarities and differences is mediated by metaphors. The recognition that many concepts in science are metaphors, that is, imported from other contexts, does not imply that these concepts are worthless or unscientific. What it does show, however, is that other concepts might also have been tried and might have been fruitful in different ways. Human knowledge is constrained by the world but also by ways of knowing, in other words, knowledge is both subjective and objective in nature.

That metaphors are at odds with the dichotomy of objectivity and subjectivity (with the latter understood as the "merely personal") has been argued at length by Lakoff and Johnson (1980; see also Lakoff 1987) in their book *Metaphors We Live By*. In this book, they discuss many widespread metaphors. War as a metaphor for argument ("Your claims are indefensible," "His criticisms were right on target") is an introductory example in this book. Such language is not "objective," but neither is it personal; the metaphor "argument is war" is part of intersubjectively shared language. Metaphors, the authors argue, undermine the dichotomy between objectivism and subjectivism, or between reason and imagination. As they involve seeing one kind of thing in terms of another, metaphors are in the domain of imagination, yet they are also in the domain of reality. The assumption of objectivism, that scientific truth and objective meaning are independent of human functioning and understanding cannot be maintained. Scientific language, like all language, is imaginative as well.

Salmon's requirement that explanation should not be "anthropomorphic" (in other words, should not be associated with human understanding) is an objectivistic requirement. For him, anthropomorphism, like pragmatism, is associated with arbitrariness and subjectivity.[3] Now it is clear that it would be highly impractical if scientific explanations differed from person to person. But the association of scientific choices with arbitrariness and subjectivity is not adequate. Not every theory is compatible with experience, not every question can be asked within a specific theory, and not every answer is true. While science is full of human images and choices, it is not arbitrarily personal.

Not only event descriptions and their explanations are linguistic entities, but so are theories. Metaphors are conspicuously present in the three approaches to development. When they are disregarded, the three approaches may appear to be "nested" in that the genetic outlook is narrower than Goodwin's organismic view, and constructionism encompasses them both in taking in the largest part of the world. This is one way of putting it, but it is in part misleading. "Programme," "field" and "construction" are the central metaphorical devices, as discussed in Chapter 2, and they shape the field of study. Genes are present in all the three approaches, but have a different conceptual role to play. In a similar way, environmental influences have a place, but again a different one, in all of them. The approaches are not simply making different selections from some complete set of causal factors. That is, they do not pick out different parts of a predefined field; instead, they define the field and its characteristics, and have competing views to offer about what is the best overall picture.

Within each picture, certain questions are near at hand and others will never arise. The genetic approach aims at clarifying in detail how genes work, Goodwin's structuralism aims at a general understanding of the physical and chemical forces determining form, constructionists aim at a better insight into the nature of interactive processes of development. Defining the subject of research clearly guides explanatory choices.

So far, we have seen unavoidable trade-offs as well as unavoidable linguistic choices as elements of a pragmatics of theories. The three approaches to development make different trade-offs, and define the field of study in different ways. Apart from that, the three rivals have their views on what it is to give a good causal explanation, and this is an important subject in the pragmatics of theories, to which I now turn.

Causal Explanation: Differences

There are many ways to think about causal explanation, and the subject has a long history. A treatment of causation with some pretension to overview the field could begin with Aristotle's four causes, go on with Hume's regularity view, and so on. I have no such goal, and neither will I be concerned with what a cause "really" is; my aim is to compare how the three approaches deal with causal explanation. In Chapter 2, I discussed how Goodwin criticized the difference-making perspective on causes that is so dominant in biology. He proposed to look at the laws of development in order to be able to explain the phenomena more completely. Different perspectives on explanation are involved. Looking for differences and looking for lawful mechanisms are strategies more or less associated with different traditions, the first one predominantly experimental and the second one aiming at theoretical understanding. Besides, the

structuralist search for lawful mechanisms shares with the constructionist approach the insistence that causation should be approached
as a relational issue. All strategies involve choices, as we shall see.

John Stuart Mill is a good starting point for the difference-
making perspective on causation, which helps to understand salience
in genetics. In *A System of Logic,* Mill says that the *real* (meaning *total*)
cause of an event is the assemblage of all the antecedent conditions.
Consequently, "the real cause is the whole of these antecedents; and
we have, philosophically speaking, no right to give the name of cause
to one of them, exclusively of the others" (Mill 1843/1973, 328).
Although in practice we pick out one or at most a few conditions
as the cause, the selection process has no theoretical basis. It is a
capricious matter, depending on purposes of the discourse.

Later philosophers have disputed the capriciousness of the selection, and all kinds of criteria have been proposed that are best
suited in practice or in theory to distinguish causes from other
conditions. For example, according to the influential analysis of causation in legal contexts by Hart and Honoré (1959), the prototype
for causal thinking is a human action that disturbs a normal flow of
events, of the type "X killed Y." Accordingly, a cause is the condition
that accounts for an abnormality, because what needs explanation, at
least in daily life, are abnormalities, and abnormal conditions count
as their cause. In this analysis the salient factor, "the difference that
makes the difference" is the abnormal factor. All the other conditions are "background conditions."

Other criteria for distinguishing the salient cause from the
inconspicuous background condition have also been proposed. They
are always defended with the help of some illuminating examples,
but none of them is universally applicable.[4] The most general conclusion that can be drawn is that somehow, the cause is "the difference that makes the difference." What exactly is the difference that
counts depends on the question that is asked.

The role of questions has been worked out in Van Fraassen's
pragmatic approach to causal explanation. Explanations, says Van
Fraassen, are answers to why-questions. Following Garfinkel and
Dorling, he analyses why-questions as involving a contrast class. When
we ask "Why P?", we really mean "Why P rather than (other mem-

bers of) X?"; where X, the contrast class, is a set of alternatives. The contrast class determines the range of potential answers to the question. It is seldom explicitly mentioned, though it is essential for understanding the question that is really asked. For example, "Why did Adam eat the apple" can be construed in various ways, with different contrast classes (Van Fraassen 1980, 127):

a. Why was it *Adam* (rather than someone else) who ate the apple?

b. Why did Adam eat the *apple* (rather than some other fruit)?

c. Why did Adam *eat* the apple (rather than doing something else with it)?

The example shows that several interpretations of the question are possible. Explanations which appear to be in conflict may thus actually be compatible because they are really answers to different questions. Van Fraassen concludes that causal explanations are context-dependent in two ways. First, the selection of the salient cause depends on a person's interests. Second, it depends on the specific contrast class that is implied in the question. As Germund Hesslow (1988, see also Hesslow 1984) has emphasized, once it is clear what question is being asked, it is often easy to understand the causal selection being made.

Causal Explanation in Genetics

In what sense are genes causes of development? The most plausible account of the notion of genetic causes is in terms of difference making, which explicates a "sober" interpretation of genetic explanation. As mentioned in Chapter 2, mutation analysis reveals what difference genetic differences make, and any genetic analysis has to start with finding or generating mutations. Fred Gifford has given an explication of the notion of cause in genetics in terms of difference: a trait is genetic (with respect to population P) if it is

genetic factors which "make the difference" between those individuals with the trait and the rest of the population (Gifford 1990, 333).
Such a causal explanation does not spell out the mechanisms that
lead to traits in individuals; not traits in individuals, but differences
between individuals have genetic causes.

For developmental genetics the difference-making perspective,
associating effects with genes through mutation analysis, is only the
beginning. Further research aims to unravel mechanisms of gene
action in molecular detail, and causal pictures become more rich and
complicated. Gifford, in the same paper, also discussed such situations
of richer knowledge: When is it justified in such situations to call
individual traits "genetic"? The relative importance of causal factors
for individual traits may perhaps be established, he argued, by a
"specificity criterion," which distinguishes between factors that
specifically influence the trait as described, and other factors with a
more general influence. For example, a protein has a genetic cause
in that "a certain genetic factor affects that protein specifically, whereas
all the other causal factors have much more general effects" (Gifford
1990, 344). Here too, the cause is a factor that makes a difference,
in this case between the trait as described and a somehow different
trait. But apparently the relative importance of causes in such cases
is at least not population dependent.

But it is, through the epigenetic conditions. Kelly Smith, in
response to Gifford, has argued that Gifford's proposal implicitly assumes that epigenetic conditions are equal. An analysis of causes can
always be made in different ways, depending on what is kept constant,
and this affects the specificity criterion as well as any other criterion.
Smith calls this the "new problem of genetics": "It seems that for any
analysis (regardless of the criterion) which reveals genes as the true
cause of a trait, there will be a complementary analysis showing the
trait to be epigenetically caused" (Smith 1992a, 347). This is true, if it
means that for any trait different explanatory questions can be asked
and unprecise questions can be made more precise in different ways;
selection of explanatory causes depends on the precise question one
tries to answer, including its implicit ceteris paribus assumptions.

Hence, whether research aims purely at associating differences
(between individuals, between proteins, or whatever) with crucial

factors or whether it goes further and looks closer at the (molecular) mechanisms involved, the identification of causes always depends on the precise question that is answered by the experiments. If environmental conditions are kept constant in order to study the action of genes, research will not yield information about the influence of different environmental conditions. In other words, when mechanisms are studied, choices have to be made as well. Below I will return to this point from a different angle.

What Is Relational Causation?

Both the structuralist and the constructionist approach, in opposing geneticism, argue for a more relational view of causation, with the intention to emphasize processes instead of entities in causal thinking. In order to clarify what is at issue, let me introduce relations in an abstract way, that is to say, relations as they appear in logic.

In logic, relations are based on the notion of the ordered pair, or more generally, the ordered n-tuple (Suppes 1957; Tarski 1965; Hodges 1978). Definitions of relations are not given in the abstract, but for specific classes, and there are two ways to do so. The first describes the relation as a polyadic predicate. The ordered n-tuples that satisfy the predicate then form the relation. The second simply lists the ordered n-tuples that form the relation. The second way evades the necessity to describe the conceptual relation between the elements of the ordered n-tuple, but becomes very impractical when the relation consists of many tuples. For example, the relation "was born before" has so many members in the domain of humans, that it is much more practical to describe the relation as the predicate just given, than to write down all the ordered pairs that form the relation.

The study of relations was initiated by Boole and De Morgan in the nineteenth century, and its conceptual importance has been noticed by many philosophers since then. Relations deserve special attention particularly against the background of their long history of nonexistence as a means of expression in logic and philosophy. In traditional Aristotelian logic, everything is expressed by monadic predicates, of the form "x is y," as in "Mary is tall." There was no place for relational expressions such as "Mary is taller than John." The Aristotelian solution was to assume that Mary and John partook in the notion of tallness in different degrees. Relations were expressed as monadic predicates somehow, which implied a loss of information in most cases, and led to many paradoxes. The introduction of relations in nineteenth century logic ended the need for such a reduction (see Barth 1974; Barth and Krabbe 1982).

Language, however, is still full of monadic predicates where relations would be more appropriate. Monadic expressions give a flavor of absoluteness to statements and neglect all kinds of embeddedness. For example, when it is said that someone is shy (where "shy" is a monadic predicate), it is too easily forgotten that shyness is shown in particular social circumstances. The dominance of monadic predicates is intimately related with a grammar that has nouns in a central position, and with an ontology that emphasizes entities and their characteristics. Excessive monadic predication is thus not an innocent characteristic of language; it is responsible for enormous amounts of context-disappearance. But doing justice to relations makes concepts more complex, and sentences more cumbersome. This partly explains why monadic predicates are so persistent: they make the world look more simple.

Hempel and Oppenheim's (1936) *Der Typusbegriff im Lichte der Neuen Logik* focused attention on some philosophically and practically important reasons for replacing monadic predicates by relations. The logic of monadic predicates is a logic of classifying concepts, generating "types." This gives unavoidable trouble in cases where differences are gradual (such as when Mary is taller than John, who is in turn taller than Paul). In such cases, concepts are needed that can generate gradual orderings. Comparative concepts like "is x-er than" (where x may be any characteristic) are relations that do the work in these cases.

But fields with gradual distinctions are not the only context in which relations matter. Comparative concepts express but one type of relations. Many other kinds exist, with widely different functions, their only similarity being that they connect two or more items. For example, relations are important in ascribing characteristics to things or persons, which is an important issue in personality psychology.

Many traits are not really monadic traits inherently belonging to a person, but traits of a person in a certain context, as is clear in cases such as shyness or dominance or passivity. Further, it matters which criterion is used to ascribe a trait; criteria also form a relevant relatum.

Since everything is related to everything else in principle, doing justice to relations can never mean accounting for all potentially relevant relations. Relations specify context, but not the whole context; they may concern some relevant relatum in the environment, or some relevant criterion or comparison, or some information about the context from which a statement is made. For example, it could be specified that John is shy in such and such circumstances, or according to such or such criterion, or that he is shyer than Mary. Other specifications, or combinations of them, are possible, but never can the total set of all possibly relevant relations be given, if only for practical reasons.

Biology, like psychology, suffers from the burden of the philosophical tradition. The issues from personality psychology in trait ascription are also relevant in the ascription of traits to animals, and cases of many different types exist. For example, it is often said that organisms are adapted, but various authors have pointed out that it makes no sense to think about adaptedness as something absolute. Wim van der Steen (1990) writes: "Things cannot be adaptive *simpliciter*. Characters of organisms can be adaptive as compared with other characters in a particular environment with respect to a particular criterion." That is, adaptednesss is a many-place predicate, not a monadic one. In the same spirit, Robert Brandon (1990) does not give a definition of adaptedness but of relative adaptedness (of one organism compared with another) within a specific environment.

Structuralist Causation

Goodwin takes a mathematical approach to relations. Mathematical equations, such as the field equations that he is looking for, represent relations between variables and parameters. The relations here have the form of mathematical operations. Infinitely many mathematical relations are possible, of widely different form. Clearly, mathematics, like symbolic logic, is a developed, flexible, and precise means of expressing relations. An emphasis on process is inherent, for example, in differential equations, which describe rates of change of a particular process in relation to a set of variables and parameters. One point of caution is that mathematical equations as such have no causal significance. Mathematical models in biology may, but need not, allow for clear-cut causal interpretations of their variables and operations. This is sometimes a source of confusion between modelers and experimental biologists.

Mathematical models, which often abstract from the concrete material basis, are increasingly successful especially in describing the generation of pattern and form. Goodwin's model for the generation of whorls in *Acetabularia* is an example I mentioned in Chapter 2. Other models in this tradition have been able to explain, among other things, the spatial patterning of teeth primordia in crocodiles (Kulesa et al. 1994), bone patterns in limbs, and stripes and spots on the skins of animals (for example, see Oster and Murray 1989).

Through such models, Goodwin is interested in finding biological laws. The word "law" has an air of generality, sternness and inescapability, expressed by the idea that nature is "governed" by laws. But it may be helpful to realize that *law of nature* is a metaphor that has political and theological origins, a point emphasized by both Nancy Cartwright and Evelyn Fox Keller (Cartwright 1983; Keller 1992, 29). This does not disqualify the concept, but reminds us that laws represent only one way of looking at explanation. The aspect of generality has especially been criticized with an eye on causation. Cartwright (1983) has pointed out that the working of natural laws tacitly rests on crucial *ceteris paribus* assumptions. In other words, they hold only in particular conditions, assumed to be constant. According to Cartwright, to think about reality as governed by "laws of nature" is not fruitful. Instead, reality is full of causal powers or capacities (some of which are described by familiar laws of nature) which through their interactions produce particular results. The result is a less generalizing way of thinking about scientific regularities, and an emphasis on the study of causal mechanisms which (unlike Salmon's) is not based on an idealization of scientific theory.

Mathematical modeling cannot do justice to all potentially relevant relations. Methodological trade-offs as well as other choices are unavoidable. Doucet and Sloep, in their book on mathematical modeling in biology, write about a specific kind of choice that model builders face: the choice of system variables, which involves crucial decisions in the analysis of complex systems. No system description can contain everything; "he [!] who wants to know everything, ends up knowing nothing" (Doucet and Sloep 1992, 306). A potentially realistic model with "some 20 state variables and 120-odd parameters" may need an enormous amount of guesswork, so that it is almost impossible to locate inaccuracies. A simple model may be more useful for many purposes; trade-offs are clearly at stake here. Doucet and Sloep add that "choice" should be taken in a wide sense; it is "not a matter of picking items from a given list, and variables must often be *created* by the modeler" (Doucet and Sloep 1992, 306; emphasis by the authors). Thus, from a mathematical point of view the conclusion is the same as from the linguistic point of view discussed earlier: fields of study are not predefined but are in some senses created by investigators.

A further important choice concerns the boundaries of the system: What does and does not belong to it? According to Doucet and Sloep, what does not belong to the model is "environment." In their example concerning growth of an animal, a variable such as food supply would normally be considered part of the environment and figure in the model as an input variable. However, it may also be taken up as a state variable, that is, a variable that is internal to the model (p. 309). In that case, it no longer belongs to the environment. Note that this use of "environment," which depends on model choices concerning state variables and input variables, is different from environment in the sense of "outside the organism," which is how environment is used throughout the present book. If a model were made of a developmental organism-environmental system in Oyama's sense, the model would include parts of the environment of the organism as state variables.

Back to Goodwin. Whether field equations represent laws or "only" regularities under certain conditions, nothing is wrong with a search for models in the form of field equations.[5] More problematic is Goodwin's view in so far as he suggests that such explanations can be complete. Field equations involve restricting choices even if they may accommodate many variables and parameters. Linguistic relativity is at issue because the relevant variables and parameters have to be identified and named. Nor can further theoretical choices be avoided, because you can never accommodate all potentially relevant factors. Even if you could, workability would decrease rapidly as more factors are taken in; model building is unavoidably subject to trade-offs and choices concerning variables. Goodwin's explanations do involve choices concerning variables, and only some of them are investigated in detail. Calcium is the prominent variable in the explanation of *Acetabularia* growth, for example. Many things are kept constant and recede into the background.

If in practice Goodwin makes choices concerning factors that are most crucial in determining changes in form, does that mean that Goodwin's explanations could also be described by the "difference-" approach to explanation? The answer is yes. In the growth and regeneration of *Acetabularia*, calcium concentrations make a crucial difference for pattern, all other things being equal. When calcium concentrations reach a critical level, symmetry-breaking processes take place in the field. Calcium is the factor that makes the difference. Indeed, the *Acetabularia* story could be told in terms of environmental calcium (see Goodwin 1994a, 86, where he comes close to doing so). However, whenever you look for mechanisms involved you hope to accomplish more than identifying the difference that makes the difference: you want to know (some aspects of) how it works. For Goodwin, the complete set of structural relations within which calcium operates is the explanatory ideal, and calcium concentration is merely a parameter value within this set, as we have seen.

Constructionist Causation

Like Goodwin, constructionists emphasize that causation should be conceptualized in a relational way, in order to represent the processes, as opposed to the entities, involved in development. As Gray expresses it: "If information and causation are our focus then it is relationships not entities that count" (Gray 1992, 194), while Gilbert Gottlieb says: "The cause of development—what makes development happen—is the relationship of the two components, not the components themselves" (Gottlieb 1992, 161–162). This should direct attention to process-aspects of development instead of static causes. In contrast to Goodwin, finding laws is not their explanatory ideal. In the constructionist perspective, interactive contingency reigns the world. "Relation," for them, is more or less synonymous with "interaction." The causal network involving genes and many other factors, including a particular environment, is what authors have in mind when they emphasize that developmental causation is relational.

This approach remains mostly in the domain of ordinary, as opposed to mathematical, language.[6] Therefore, the means of expression are those of ordinary language. As the structure of most languages invites emphasis on things and their characteristics, doing conceptual and experimental justice to the complexities of relations is not simple. Attempts to make language more suitable for describing processes may end up, for example, in the almost intractable idiom of Whitehead's (1929) *Process and Reality*, which contains a host of new concepts and words that have hardly found their way into normal use. But however difficult it may be to influence language, new adjustments of language and thought are constantly proposed. For example, people who argue for a relational view of development frequently use the prefix "co-", in terms such as co-action, co-determination, co-causation or co-regulation.

Constructionist emphasis of the environment is found in psychology as well as in biology. For example, Alan Fogel (1993), who defends a relational approach to human development, associates "relation" with "communication." The traditional subject matter of Western psychology and philosophy, he says, is the individual. In such a perspective, social processes may modify the faculties of persons, but essentially, these faculties are self-contained boxes. Relationships, in this approach, are conceptualized only as inputs and outputs of the individual psyche. Fogel argues that it is necessary to take the opposite view: "The essential fact about organisms is not their organic integrity but their connectedness to the environment". He advocates the metaphor of "co-regulated communication process" as a starting point for describing development (Fogel 1993, 60-61).

Similarities and Differences

Relations in the formal sense and relations in the ordinary language sense of "causal interaction" do not coincide. A formal (logical or mathematical) approach is flexible and not necessarily causal. Ordinary language is more heavily bound by traditional meanings, connotations and grammatical constructions. Evidently, this cannot imply that ordinary language should be abolished; science is unthinkable without ordinary language. But the respective strengths and weaknesses of ordinary and formal language approaches can be a source of polarisation. For example, Depew and Weber describe defenses of narrativism, or story-telling, in recent Darwinism. Gould and Lewontin are well-known for their rejection of adaptationist stories, but they do not reject story-telling as such; indeed, Gould is a great advocate of narrative in biology. As Depew and Weber emphasize, narrativism in biology is one way of acknowledging complexity (Depew and Weber 1995, 390). But from the perspective of those biologists who would like biology to be an exact science, the narrativist turn "can very quickly become an object of ridicule" (p. 398).

Polarization is probably not the most fruitful relation; mutual information and adjustment is more constructive. At the very least, both a linguistic and a formal approach to causal relations lead to the conclusion that completeness is unattainable. The guiding idea in the formal approach is that relations simply establish some (any) simple or complex relation between two or more items. The linguistic approach inherently involves meaning. From both perspectives it is easy to see that there are many different ways to express relations between different aspects of processes. At the same time, it is clearly impossible to establish a relation between all the relevant relata.

In analogy to monadic predication of the form "x is y," we could say that "monadic causation" is of the form "y causes x." If we think that this is in many cases a very unsatisfactory way of causal explanation, there are many ways open to make the account more relational. The following two different ways, involving difference-making and study of mechanisms, respectively, are relevant in the present context of discussion.

1. The difference-approach. A "monadic" way of expressing genetic causation is the "gene for x"-talk. In fact, genes do not cause traits; genetic differences cause differences in phenotypic outcome. As Wim van der Steen expresses it: "We should not construe 'genetic determination' as a one-place predicate but as a three-place predicate. For example, a difference in a feature between two organisms can be genetically determined in the sense that it results from a genetic difference between them" (Van der Steen, 1993a, 27). This, indeed, is one way of making causation more explicitly relational: making causes as well as effects more specific (differences rather than entities). Through variation of one factor, while everything else is kept constant, it can be studied what difference that factor makes.

2. Study of mechanisms. When the subject of explanation is not a difference, but a trait, for example, relations can be introduced in a different way: by describing the process or mechanisms involved. Since the cause of a trait cannot be an entity, but must always be a complex process in which different entities play roles, a developmental outcome is caused by a network of interacting entities and processes. This network is extremely complex, and in studying it one will have to focus on particular parts of it, keeping other things constant.

The restrictedness of causal accounts is more visible in the first approach, but inevitably present in both.

Back to the whole picture. As discussed in Chapter 2, the three approaches to development differ in their explanatory ideals, but they are not incompatibly, unintegratably, absolutely different. The upshot of the preceding discussion about causation is that the differences between the three approaches in how they view causal explanation should be put in perspective. In the first place, genetic accounts of causation, when expressed and interpreted rightly, are just as relational as structuralist and constructionist accounts. In the second place, no relational causal treatment of x can be complete about the causes of x, and this applies just as well to the structuralist approach as to the other two.

While causal choices are inevitable, the approaches differ in the kinds of choices they make, as we have seen. Yet when it comes to content, points of similarity and overlap also exist.

For example, think of gene regulation. Constructionists as well as developmental geneticists are interested, in principle at least, in environmental gene regulation. Their different overall perspectives may make them describe such phenomena in different terms (from a constructionist point of view, "gene regulation" may sound uncomfortably gene-centric to begin with) and they may also differ on how to proceed once some environmental influence has been found, but the mechanisms of gene-environment interaction are interesting from both perspectives.[7] "Epigenetics," the study of the regulation of gene expression, is a domain where geneticists and constructionists may meet. This point will return in Chapter 5.

Similarly, both geneticists and structuralists are interested in questions concerning embryonic pattern formation and the role of gradients in this, different though they deal with the gradients fom a theoretical point of view. For geneticists, the gradient will primarily have a function in differential gene regulation, while a structuralist sees it as a field-determining element. In fact, the theoretical situation concerning gradients, and pattern formation in general, is even much more complex. Lewis Held (1992) has made a classification of models for embryonic periodicity. His monograph illustrates that fields, gradients, waves, progress zones, and other structuring devices can play many different theoretical roles. He distinguishes no less than six categories of models to explain pattern formation, containing a total of nineteen different specific models. His book also illustrates that developmental biology can be divided in many ways. A lot of research is going on and the dividing lines that I have drawn are not relevant for all purposes.

Metaphors too may be a source of narrowing rather than widening gaps. The central metaphors that are associated with the three views emphasize the difference between them, but there are many local metaphors which are not so laden. Take the use of conversation metaphors, which constructionists as well as geneticists make use of. The general tendency is for geneticists to suggest that genes give instructions, and for constructionists to portray development as a dialogue between equals. But the metaphor can be used in more than these two ways. For instance, much research is done on cell-to-cell signaling in development. Research on signaling pathways may well give new directions to the views of development. In descriptions of this research, conversation metaphors are prominent, and though the theoretical framework is often a genetic one, in descriptions of the conversation going on between cells the genes are often only background actors.

The overall views have thus points of contact and overlap. Nevertheless, it remains true that they give different heuristic advice on what is most important to understand about development, and thus implicitly also on what is not so important to know. Their advice reflects the specific strengths and limitations of the approaches, which are no problem as long as they are recognized. The idea of trade-offs emphasizes that such restrictions cannot be avoided anyway.

Implications for the Pragmatics of Causal Explanation

The foregoing has argued and illustrated that scientific explanations should be understood in the context of their overall theoretical approach. Explanations are affected by the choices that characterize such approaches and they can only be "complete" within this specific context.[8]

In the beginning of the chapter, I argued that a pragmatic philosophy of science should also be pragmatic about its own tools. While this in itself is a reason to reflect upon the goals and effects of the usual pragmatic approach, there are also more specific reasons to look critically at it: a) its uniform reconstructions of explanations, and b) its neglect of the theoretical context. Let me deal with these issues separately.

First, uniformity of reconstruction. In Van Fraassen's and Hesslow's perspective on causal explanation, an explanation is an answer to a why-question about a specific difference. This question approach to explanation is often regarded as *the* pragmatic approach to explanation. If it is characteristic of a pragmatic approach that there is no absolute way of analyzing anything (see Pepper 1942), this identification is misguided. In order to see how different explanations relate to each other, it is certainly often helpful to see explanations as answers to questions, and to wonder what exactly the question is that a specific explanation does answer. But first, this focus on questions is just one tool to understand explanations, and second, if this tool is used, why should all questions be understood or reconstructed as having the same form? There is no need to make universal assumptions of what an explanatory question is or should be like. Specifically, an explanatory question is not necessarily a why-question, and it need not always ask for a difference. Let me explain both points.

1. Van Fraassen's restriction to "why"-questions is not adequate, unless you feel able to reconstruct all "what" and "how" questions as why-questions. Such what and how questions can also be explanation seeking (see also Von Eckardt 1992, 21). Matti Sintonen, too, acknowledges such questions, though he sees them as special kinds of why-questions (Sintonen 1993).

2. The heavy emphasis on difference making, which is implied in reconstructions of the form "Why X rather than . . ." is not necessary. Any question can be reconstructed that way, but it can also be helpful to acknowledge that some questions are meant to ask for mechanisms. Explanations that use other notions of cause than the difference-notion can be considered as answers to questions as well. Goodwin's field equations, discussed previously, are an example; they are meant to describe (parts of)

processes. It may often be more illuminating to reconstruct such questions as mechanism-seeking than as difference-seeking.

Second, the pragmatic analysis of scientific questions needs an extension by placing questions in their theoretical context. Both Van Fraassen and Hesslow refer to subjective factors or personal interests that determine questions. Personal interests determine the actual structure of explanatory questions. But questions are not only asked by a person, but also in a specific conceptual, historical, etc., context. In science, the theoretical context is often more important than personal interests.

That scientific causal questions have a theoretical context is not a new or surprising point. It has been stressed decades ago by Hanson, who writes that "what we refer to as 'causes' are theory-loaded from beginning to end" (Hanson 1958, 54). But in Van Fraassen's treatment of explanation, the theoretical context is an idealized background and cannot illuminate theoretical backgrounds of conflicting explanations.[9] The result of this idealized and single context, in combination with the particular form of question-reconstruction, is that pragmatics has come to look like a strategy for defusing conflict: it typically shows that apparently conflicting explanations are actually answers to different questions and are thus not really in conflict.

Introducing the theoretical context changes this picture; scientific questions may also give rise to different answers because they are posed in different theoretical contexts. When two people come up with different answers to the question "Which are the causes of development?," their theoretical orientation may make them answer this question differently.

The importance of theoretical context for causal explanation in biology has been discussed by Sandra Mitchell in a paper on pluralism and competition in evolutionary explanations (1992). Mitchell asks when and how there can be a real conflict between explanations in biology. Sometimes, she says, people will apparently give different answers to the same question, but really ask different questions. This is a situation in which a pragmatic approach is illuminating. In other cases, people will agree about the question but give different answers. In these cases, not a classification of questions but of answers is needed and this brings her to the role of theory. In her view this implies leaving a pragmatic approach behind since a pragmatic approach can only deal with (resolve) conflicts that are really about different questions. This is right with respect to the usual pragmatic approach, but not with the widened pragmatic approach that I propose.

Mitchell proposes to see conflicting answers as appeals to distinct abstract models of causal processes, which concentrate on some causes and abstract away others. As she sees it, their integration on a general level would be accomplished in a theory including all potential causal factors. She is rather skeptical about such integration for the cases she discusses.[10] "There may be some reason to believe that a single formulation will not be found to adequately represent all contributing causal factors" (Mitchell 1992, 143). As I have argued earlier in this chapter, the difficulty of finding those "single formulations" has its basis in the nonexistence of a predefined set of causes, which is reason to doubt unified theory in theory as well as in practice. Conflicting approaches define the field in rival ways, so that the "same" factor has a different role or is differently defined in each of them.

Therefore, when questions are considered in relation to their theoretical background, question explication is not necessarily a strategy that leads to the defusing of conflicts; it can also be a means to locate questions in their respective theoretical backgrounds, and thus, to may make conflicts more visible. Such theoretical conflicts may be resolvable, but question explication will not do the job.

A plurality of scientific approaches is no sign of weakness; on the contrary, it is often very fruitful. It prevents any approach from becoming too self-evident. When a specific kind of explanation figures as *the* explanation, and is tacitly assumed to be helpful for all purposes, this is much more of a problem. The genetic approach is now so dominant that in many contexts this problem applies to it.

4 ✐ Development and Evolution

Development and evolution have been separate disciplines for a long time, but they should finally be integrated, according to an increasing number of biologists. Not surprisingly, it is in part controversial what the integration should look like. In this chapter, I concentrate on conceptual devices that influence the relationship between the disciplines. The proximate–ultimate distinction has played an important role in keeping development and evolution apart, but is losing influence now that the disciplines are being integrated. In this integration, another distinction is very influential, however, which is the distinction between internally caused development and externally caused evolution. The distinction discourages acknowledgment of external influences in development; it suggests that the only causal role of the environment is to select among organisms. An integration based on this distinction thus reinforces emphasis on internal causes in development. Another issue in this chapter is the tendency of evolutionary considerations to overshadow developmental ones. Since causes of development deserve attention in their own right, there are good reasons not to blur the boundaries between evolution and development completely, I will argue.

Integrating Separate Disciplines

In the beginning of the twentieth century, the study of heredity involved transmission as well as development of hereditary characteristics. In other words, genetics and embryology were integrated in a single field. Evolution was a different matter; it had an uncertain status, and for many, it did not even belong to biology proper. But the relations between these fields underwent important changes in the decades that followed. By the 1930s, the definition of heredity had become narrower. It now involved only transmission, and embryology no longer fitted in. Genetics and embryology had thus become separate disciplines. Next, genetics was involved in the evolutionary synthesis of the 1930s, at first predominantly in the form of population genetics.

In a paper about the split between embryology and genetics, Garland Allen (1985) has documented how the change took place in its leading architect, T. H. Morgan. Morgan, who had been trained as an embryologist, worked within the integrated framework at first. In 1910 he wrote that "we have come to look upon the problem of heredity as identical to the problem of development. The word heredity stands for these properties of the germ cells that find their expression in the developing and developed organism" (Allen 1985, 113–114). In the embryological definition of heredity, transmission was just one part of the larger problem how offspring come to resemble their parents.

By 1926, however, Morgan had redefined heredity as the study of transmission and had divorced this field of research from embryology, the study of development. According to Allen, an important issue behind this change was the growth and spread of mechanistic materialism in biology, as advanced by Johannsen, among others. Instrumental in Morgan's change was Johannsen's distinction between genotype and phenotype. *Phenotype* was a derogatory word, which referred to the morphological and descriptive view of heredity of the old natural history tradition. Mendelian genetics should be rigorously separated from that old tradition, and Johannsen thus gave a new meaning to the term heredity (Allen 1985, 127). Although it is not certain, says Allen, whether Morgan actually knew Johannsen's work, the distinction was very appropriate to help him move from his earlier to his later views. More generally, the prevailing analytical methods favored by the mechanistic-materialistic school of thought made the climate ready for the separation.

The gap between the disciplines remained wide for quite a time, and the outlook in both disciplines differed considerably. Klaus

Sander (1985, 364; see also Hamburger 1980) has remarked that for a geneticist, a cell was primarily a nucleus, while embryologists emphasized the cytoplasm. Genetics evolved as a reductionist discipline while embryologists favored holistic approaches.

Embryology became separate not only from genetics; neither was it integrated in the synthetic theory of evolution framed in the 1930s. In fact, the separation between evolution and embryology was even older; it had its origin in the opposition of embryologists to Haeckel's biogenetic speculations. At the end of the nineteenth century, first His, and later Roux favored the study of the immediate, instead of the phyletic, causes of development, in opposition to Haeckel and his biogenetic law. They advocated research into the 'mechanics of development' (Entwicklungsmechanik). As Gould points out in *Ontogeny and Phylogeny* (1977),[1] His readily admitted that his preferred mechanistic explanations did not exhaust the content of causality. The immediate forces responsible for development also had a phyletic origin; mechanistic and phyletic explanations could thus exist alongside each other. Yet, says Gould, he found the mechanistic causes both more significant and more modern because of their conceptual link to the mechanistic physiology of his time, and he said: "To think that heredity will build organisms without mechanical means is a piece of unscientific mysticism" (His quoted by Gould 1977, 191). His wrote in 1874; around 1890, Roux launched a similar offensive and with more success. Entwicklungsmechanik triumphed and "set the fashion for the next half-century, one of the most exciting and fruitful periods in the history of embryology" (Gould 1977, 194).

Gould uses the terms proximate and ultimate, introduced in 1896 by Romanes (Gould 1977, 188) to describe the different explanatory aims: the causes sought by Entwicklungsmechanik were proximate, while ultimate causes are to be found in natural selection. The distinction between ultimate and proximate causes has been widely used by evolutionary biologists to point out how far apart embryology and evolution are. Ernst Mayr (1961), in his influential paper "Cause and Effect in Biology" identifies proximate causes with immediate causes, acting during an organism's lifetime, while ultimate causes are causes with a history that have been incorporated into the system through natural selection. Later, he changed "ulti-mate" into "evolutionary," so that the distinction was now between proximate and evolutionary causes.

In the terminology of proximate *versus* evolutionary, the dis-tinction virtually implies that development is irrelevant for evolution. Indeed, for Mayr it serves this function. Evolution has to do with changes in the genetic program, development with the decoding of the program, which plays no role in evolution.

However, a widespread dissatisfaction has grown in biology with this separate status of embryology. Integration of evolution and development is now seen as necessary by many, from different quar-ters. One idea invariably involved is that development is more impor-tant for evolution than has traditionally been thought.

In the first place, this is a corollary of the evolutionary impor-tance of the phenotype, as Lewontin has argued in *The Genetic Basis of Evolutionary Change.* The topic of this book is population genetics, the discipline which has long been regarded as the core of evolu-tionary biology. Population genetics at first sight appears to be a theory of genotypes only, says Lewontin, but since it includes fitness as a parameter, which is a function not of the genotype but of the phenotype, population genetics inevitably concerns the domain of phenotypes, too (Lewontin 1974a, 15).

Second, not only are embryological stages increasingly seen as full-fledged stages in life histories, for example in terms of their energy budgets (see Kooijman 1993, 49); development is also in-creasingly regarded as a source of evolutionary forces in its own right. Embryological patterns are centrally important in defining phylogenetic groups (on the level of phyla and higher) with different basic organizational structures or "Baupläne." Questions concerning the evolutionary relations between these different Baupläne and the role of developmental constraints and/or epigenetics are increasingly asked (see Willmer 1990; Hall 1992; Williamson 1992; Jeffery and Swalla 1992; Slack et al. 1993; Rollo 1994; Horder 1994).

This much is perhaps uncontroversial. But ideas vary on exactly which lessons should be learned from development and on how the integration should take shape. Such views are associated, not surpris-ingly, with perspectives on the disciplines to be integrated, developmental biology, and evolutionary biology. Thus, the three views distinguished in Chapter 2 entail different views of the integration of development and evolution.

Those who think that development is controlled by genes evidently see genetics as the central discipline in the integration. Raff and Kaufman (1991) in their book *Embryos, Genes and Evolution* see evolution as change in genetic programs of development. In the same spirit, Scott Gilbert ends the fourth edition of his textbook *Developmental Biology* by announcing a new unified theory of evolution, brought about by the coming together of developmental genetics and neo-Darwinism (Gilbert 1994, 857).

Critics of geneticism also see a close connection between development and evolution, but not with genetics at the center. Their views on integration are implicitly present in Chapter 2. In Goodwin's view, evolution should be seen as a constrained journey through the set of forms that are possible from a physico-chemical point of view. According to Oyama, evolution is change in organism-environment systems, with environmental factors and genes in equally important roles. Their specific views of integration will become clear below through the discussion of relevant concepts and distinctions.

Proximate and Ultimate Causes

Several concepts and distinctions influence views of the relations between evolution and development, and I shall analyze the form these influences take in the three approaches. Let me first of all return to the distinction that has kept the disciplines separate for so long, the distinction between proximate and ultimate causation. All approaches to an integration of evolution and development share some unhappiness about it.

Mayr did not invent the distinction between proximate and ultimate causes, but he did make it prominent in biology. According to him, a distinction can be made between the immediate or proximate causes of phenomena, such as physiological causes, which figure in answers to how-questions, and ultimate causes which are answers to why-questions, that is, evolutionary questions (Mayr 1961). In later characterizations, proximate causes relate to "all aspects of the decoding of the information contained in the DNA program of the fertilized zygote," while ultimate causes, now called evolutionary causes, concern "the laws that control the changes of these programs from generation to generation" (Mayr 1976; 1993).

Since the proximate-ultimate distinction serves the purpose of keeping development and evolution apart, there is naturally a growing uneasiness about it in this time of integration.[2] Mayr, however, has always remained a fervent proponent of keeping developmental and evolutionary causes apart on the basis of the distinction. In 1993, during a symposium in honor of him, he said: "I must have read in the last two years four or five papers and one book on development and evolution. Now development, the decoding of the genetic program, is clearly a matter of proximate causations. Evolution, equally clearly, is a matter of evolutionary causations. And yet, in all these papers and that book the two kinds of causations were hopelessly mixed up" (Mayr 1994).

However, the distinction itself is a source of some confusion. One reason might be that Mayr has used it in different contexts in different phases of his career, and for different purposes, as John Beatty has documented. From the 1950s onward, when molecular biology was gaining prestige, he used it to defend the importance of evolutionary studies and to "make the point that there is more to biology than the study of proximate causes" (Beatty 1994, 349). In such a context it is easy to imagine that "ultimate" gets the connotation of "more fundamental," while at other times it seems to mean simply "more remote in time."

Another source of confusion is that the proximate-ultimate distinction is characterized on the one hand by reference to different questions, and on the other hand by the causes that figure in the answers to those questions. Thus, on the one hand, questions concerning development yield proximate causes, while questions about evolution yield ultimate causes. On the other hand, it is suggested that there are two disjunct sets of causes, the one proximate and the other ultimate. But these characterizations do not necessarily coincide. The existence of different kinds of questions does not imply that a given causal factor can figure in the answer to only one of them. A cause of development may well be a cause of evolution at the same time; "developmental constraints" are such causes.

As I will argue later in this chapter, it is often useful to distinguish between questions about evolution and questions about development. But the distinction between proximate and ultimate causes is not a great help. If it refers only to "answers to questions about ontogeny" and "answers to questions about phylogeny" it is superfluous. If it refers to something more, such as different sets of causes, it can only be a source of confusion. Any integration of development and evolution has to clarify those confusions somehow.

Internal versus External Causes

Another distinction that has made ontogenetic and evolutionary development look quite different is the one between internally caused development and externally caused evolution. It is very influential in the present integration of development and evolution.

Lewontin (1983b; see also Levins and Lewontin 1985, 85–106) and Sober (1985) advanced the distinction between a transformational and a variational model of change, in order to elucidate what was new in Darwin's approach to biological change. Lamarck had used for his ideas on evolution the older model of biological change, which is the developmental, transformational one. Here, evolutionary change is ontogenetic; it originates in the development or transformation of organisms. Darwin introduced a new model of evolution. Variations arise at random and are then selected. Evolutionary change takes place not within but among organisms, at the population level, by the selection of varieties, so that the composition of populations changes. The development of the different varieties is irrelevant; it is only their (static) differences that matter. Darwin thus replaced a view of evolution centering on individual development with a selection view working at the population level, in which individual development played no role.

But this is not all. In characterizing the two conceptions of change, Lewontin and Sober both emphasize that in the transformational model causation is from within: development is an unfolding that is governed by inside forces. In the variation/selection model of change on the other hand, change is governed by the outside force of selection.

Thus, the distinction is a double one; it not only distinguishes between individual and populational models of change, but it also contains the idea that individual development is caused by internal forces, while change on the population level is caused by external forces. Lewontin criticizes classical Darwinism precisely for containing this assumption that "the genes propose and the environment disposes" (Levins and Lewontin 1985, 88). In his view (see Chapter 2), development is not an unrolling, and the environment is not an autonomous selecting force; the organism and the environment actively codetermine each other.

On the basis of the same view, Susan Oyama has criticized Sober and Lewontin's characterization of the transformational/variational distinction, in so far as this characterization suggests that the combination of the two different elements of the distinction is necessary. (Oyama 1988, 1992b). This is not so, she points out: individual change can be decoupled from the idea of predetermination, and change in populations does not imply that the variants within the population are static. When these connections are dropped, the transformational model and variational model are no longer in opposition (Oyama 1988, 257).

Oyama (like Lewontin) argues that evolution is a process of dynamic, not static, entities. When developing organisms and their environments are seen as developmental systems, this implies that what is transmitted to future generations is not just genes, but other developmental resources as well. Evolution is change, not necessarily in gene frequencies, but more generally in the constitution and distribution of developmental systems (including genes). Thus, a different picture of development goes hand in hand with a different picture of evolution. Both involve, Oyama stresses, a view of change that is systemic and interactive. She concludes that "if development is to reenter evolutionary theory, from which it has long been excluded, it should be a development that integrates genes into organisms and organisms into the many levels of the environment that enter into their ontogenetic construction" (Oyama 1992b, 227).

Integration: Internal Constraints?

In spite of the shortcomings of the transformational-variational distinction as it is presented by Lewontin and Sober, it involves a historically accurate description of dominant Darwinian views, which continue to be influential. The particular kind of integration of evolution and development that is encouraged on this basis preserves the shortcomings. When development is regarded as being caused from within, the reintegration of development in evolutionary theory comes down to an inclusion of internal causes in evolution. This is precisely what happens in both structuralist and neo-Darwinian views. S. J. Gould, for example, who writes that nothing in his whole career has given him more pleasure than watching this re-integration, adds that it has become possible, now that the genetics of development are unlocked and genetic developmental hierarchies are uncovered. What is happening is a turn to the internal causes of evolution: "Nothing in the strict Darwinian paradigm suggests hostility to developmental issues, but the theory's central logic offers precious little space for the major internalist and structuralist themes of embryology either. Darwinism is fundamentally an externalist theory of adaptation, step by insensible step, to changing local environments" (Gould 1992, 276–277).

Gould is not alone in this evaluation. For example, McKinney and McNamara, in their book on heterochrony, likewise see the evolutionary turn to development as a turn to internal (intrinsic) causes: "The momentum that is being generated by papers written in the last decade has the potential to return the study of intrinsic factors in evolution back to their rightful place in evolutionary theory" (McKinney and McNamara 1991, 12).

"Constraint" is a much used notion in this context. According to Gould, the most powerful constraint on evolution is development itself: development imposes internal constraints on evolution. This view was launched in the classic paper "The Spandrels of San Marco" (1979), where Gould argued, with Lewontin, against a "Panglossian paradigm" that understands evolutionary outcomes purely in terms of adaptation. Against this view, Gould and Lewontin argue that "basic body plans of organisms are so integrated and so replete with constraints upon adaptation that conventional styles of selective arguments can explain little of interest about them" (1979, 594). According to them, developmental constraints "may hold the most powerful rein of all over evolutionary pathways." Since then, Gould has remained a spokesman for this view, interpreting developmental constraints as internal constraints.

The concept of constraint has been widely used in writings about integration of development into evolutionary theory. The term is often used in the colloquial sense that it is some kind of restriction or restricting factor, but its meaning has also been the subject of discussion. Influential is Gould's (1989) view that constraints are those causal factors which are not part of the theory you work with. Hence, the nature of constraints is not fixed but depends on the theoretical context. As soon as they are integrated in prevailing theory, they are no longer constraints, according to this definition.[3]

Geneticist and structuralist integrators share the idea that developmental constraints on evolution are internal constraints. But behind this shared notion of constraint, important differences are nevertheless hidden. Ron Amundson (1994) has argued to this effect, starting from the definition given in the multiauthored review "Developmental Constraints and Evolution." Developmental constraints are defined in that paper as "biases on the production of variant phenotypes or limitations on phenotypic variability caused by the structure, character, composition, or dynamics of the developmental system" (Maynard Smith et al. 1985). This definition, says Amundson, leaves room for two different uses that vary considerably in the importance they attribute to processes of development. The point is: What is it that is being constrained? Here, different views of evolution become relevant. Neo-Darwinians are thinking about constraints on adaptation. Such constraints are contingent obstacles, internal or external, developmental or otherwise, which natural selection cannot overcome. Developmentalists on the other hand are talking about constraints on form. According to them, constraints limit the forms that are possible in morphospace. They want to know the generative reasons why not all imaginable forms are realized, the explanandum being form, not adaptation. This is the structuralist program, with Goodwin as one representative. (According to Gould's proposal, of course, these generative reasons cease to be constraints as soon as they become part of your theory.)

Amundson's comments are illuminating in that they highlight some consequences of the differences between neo-Darwinians and structuralists. Mentioning constraints will not automatically direct attention to development: when you are interested in constraints on adaptation, the black box of development may largely remain closed. However, as he equates "developmentalism" with "structuralism," the approaches he deals with share the assumption that development is caused from the inside, and that developmental constraints on evolution are thus internal constraints. The constructionist criticism, which points out the inadequacy of the internalist assumptions about development, misses from this analysis.

Internalism is also prominent in the work of Stuart Kauffman, a structuralist thinker although his work differs considerably from Goodwin's. Kauffman has not only written papers with titles such as "Developmental constraints: Internal factors in evolution" (Kauffman 1983), but also the book *Origins of Order*. In this book, he raises the question what the origin might be of order in the biological world, and gives the answer "mainly self-organization," as opposed to "mainly selection" (Kauffman 1993). Selection as the main origin of biological order is the target of Kauffman's criticism: it has a limited role and cannot be the most important explanation of order. The explanation must rather be found in ahistorical principles of the organization of life. While in this respect his ideas are similar to Goodwin's—a similarity which they both emphasize—they differ in how they see the role of genes. Goodwin downplays the causal and conceptual role of genes and argues against the notion of genetic program. Kauffman on the other hand makes genes central. Although, like Goodwin, he opposes the current notion of genetic program, he proposes not to do away with it, but to give it a different sense. For Kauffman, a genetic program is a Boolean network.

Kauffman's book contains a message based on results of thirty years of computer simulation with Boolean networks; such networks contain a good deal of spontaneous order. The biological interpretation is that the genes of a cell can be seen as a Boolean network. A genetic program is therefore a parallel working network, with all the genes changing simultaneously. A cell type can be seen as an attractor or end state of the network. In ontogenetic differentiation external perturbations make cells go from one attractor to the next. Such a perturbation may consist of an asymmetric distribution of gene products in daughter cells, or of an exogenous influence. Kauffman offers this view as an alternative for the hierarchical idea of a genetic program with master genes that regulate genes lower in the hierarchy. At the same time, it is an alternative for a reductionist, molecular approach that aims at understanding through details. Kauffman emphasizes that his is an "ensemble" theory, that seeks understanding not by investigating details but by laying down the general characteristics of living organisms, given the nature of basic processes.

Boolean networks are abstract things allowing different biological interpretations. In Kauffman's book, they consist of genes. The "self" of self-organization thus refers to genetic networks. This is an internalist picture of development. Outside influences are not forbidden, they are even acknowledged as influencing networks, but nevertheless Boolean networks are self-organizing. Oyama has therefore pointed out that Kauffman opposes externalist explanations in terms of selection with internalist ones, thus maintaining the internal-external opposition as conceptually central. It would be much better to change the conceptual framework that demands such oppositions: "We must alter the very ground on which opponents take their positions" (Oyama 1992a, 230).

Unavoidably, when development is seen as coming from within, integrating development into evolution amounts to the emancipation of internal causes in evolution. This applies to geneticism and structuralism alike, however different their views on constraints are.

Different Roles for the Environment

Apart from influencing the integration of development and evolution, the distinction between internally caused development and externally caused evolution also affects the study of environmental influences. It implies that there is no conceptual place for them in the study of development, but only in the study of evolution. No wonder perhaps that environmental influences in development are often studied from an evolutionary perspective. A typical question is in which kind of environments the use of environmental triggers in development will be adaptive. In other words, the question is, Under which conditions does the dependence on environmental influences fit in with evolutionary theory? This is a legitimate question, but it is not about developmental process.

Moreover, when there is no conceptual space for a direct causal role of the environment in development, environmental influences in development are unavoidably surrounded by confusions. The developmental and the selective role of the environment somehow come together, for example, in the concept of trigger, which has ontogenetic as well as phylogenetic meaning. Ontogenetically, an environmental factor triggers some part of the genetic developmental program; phylogenetically, the genetic program has been adapted to use the trigger. As discussed in Chapter 2, the environmental trigger is not considered as a real cause of development; it is a "mere" trigger.

Only if the dichotomy between internal development and external evolution is dispensed with can external influences on development become plainly visible. This implies loosening the links between development and internal causes as well as between external causes and evolution.

Distinguishing between selective and direct causal *roles* for the environment is not the same as distinguishing selective and developmental environments. Robert Brandon (1990) has argued against the latter distinction precisely because the same environment may have different roles. I agree, and I add that it does make sense to emphasize these different roles, especially from the developmental point of view. As discussed in the foregoing sections, the developmental role of the environment has long been conceptually absent, which has resulted in environmentless genetic accounts of development. For the purpose of creating a conceptual space for direct causal environmental influences in development, the distinction between a selective and a developmental role for the environment can be helpful. Chapter 5 will be devoted to examples of such direct influences.

Epigenetics and Evolution

When the developmental role of the environment is not acknowledged, neo-Lamarckism may seem to be the only alternative
to neo-Darwinism. In the former, the environment directs mutations
while in the latter it selects among random mutations. If the emphasis is on these alternatives, the controversy surrounding the integration of development and evolution centers on the role of selection
(see for example J. W. Atkinson 1992a). Adding a causal role for the
environment in development may undermine this dichotomy and
help shift the locus of attention, not only concerning development
but also in discussions on evolution.

Environmental influences in development do have evolutionary
relevance. This has been stressed by Johnston and Gottlieb in their
paper "Neophenogenesis," in which they argue against defining
evolution in terms of gene frequencies, and in favor of making the
phenotype more central. According to them, this requires a theory
that involves "all of the mechanisms that may produce phenotypic
change." In particular, such a theory should explain the relationship
between genetic and extra-genetic sources of evolutionary change
(Johnston and Gottlieb 1990, 474). They favor an approach to development that is much like Oyama's, in which genetic as well as
environmental influences have a place. Like Oyama, they deny the
usefulness of the distinction between inherited (genetic) and acquired traits, arguing that all traits are both: all traits need genetic as
well as environmental influences to develop. Dropping this distinction does not require the postulation of "genetic assimilation," viz it
does not require nongenetic characteristics to become genetic; they
emphasize that their proposal is not neo-Lamarckian.[4] Instead, environmental conditions should be seen, each generation anew, as part
of the ontogenetic process. When phenotypes change in a relatively
enduring, transgenerational way through non-genetic means, such as
a different climate or new habits, evolution has occurred. For example, dietary changes in rodents (such that parents offer new kinds
of food to their young) may lead to evolutionary change in jaw
morphology, because bone growth is partly influenced by mechanical forces (Johnston and Gottlieb 1990, 478). The environment is

thus not a source of anti-Darwinian evolution in competition with selection, but influences evolution through development.

Recognition for the evolutionary importance of epigenetic mechanisms, including environmental influences, is growing. For example, it is shared by Brian Hall's and C. David Rollo's recent integrative treatments of development and evolution. In his book *Evolutionary Developmental Biology* Hall (1992; see also Hall 1983) refers to cases where predators influence the development of their prey via substances released into the environment, which illustrates that epigenetic interactions can include species-to-species induction (Hall 1992, 127). Hall stresses the consequences for understanding evolutionary change and advocates the study of "the inter- and intraspecific causal links between inductive changes in embryonic development, ecological adaptation and evolutionary change" (p. 129). Environment-including epigenetics is thus an important source of evolutionary change; new habitats may be initiators of evolutionary change through their epigenetic effects.[5] Rollo's *Phenotypes* (1994) also stresses the importance for evolution of a regulatory view of the genome and an epigenetic approach to development that includes phenotypic plasticity. Certainly, increasing insight into molecular epigenetic mechanisms gives increasing plausibility to these views.

Jablonka and Lamb (1989) have proposed a model for "The Inheritance of Acquired Epigenetic Variations" (title of the paper). Gene expression is importantly influenced by the three dimensional structure of DNA and proteins called chromatin, which can be called the gene's phenotype. Environmental stimuli can induce changes in chromatin structure, and this state can be propagated during cell divisions. Since the Weismannian doctrine of the universal separation of the germ line has convincingly been undermined (see Buss 1987), this implies that in many species environmental influences can be passed on to the next generation through the gene's phenotype. This phenomenon may be widespread in plants and fungi, and more restricted though still present in organisms that do have their germ lines separated from the somatic cells. Jablonka and Lamb see heritable epigenetic changes as "Lamarckian" mechanisms, but note that no change in DNA base sequence is involved. In a later paper, Jablonka et al. (1992) have discussed the importance of epigenetic inheritance for evolution. They argue that epigenetic inheritance may have played a large role in the evolution of multicellularity.

Developmental Systems and the Boundary Problem

The traditional distinctions between development and evolution go with a one-sided internal picture of developmental causation, and it is clear that an integration of development and evolution will not automatically correct this. Indeed, an integration focusing on development adding 'internal constraints' to evolution only continues the one-sidedness. Only in a constructionist or at least epigenetic perspective, the environment can be a cause in both development and evolution. But even in this perspective, I will finally argue, the integration of development and evolution is not automatically an unmixed blessing because it may hamper attention to the causes of development.

Various constructionist authors have argued that both evolution and development are dynamic interactive processes. For them, the step from talking about evolution to talking about development and back is often very small. Sometimes, the distinction between the two almost disappears. Gray, for example, switches freely between developmental and evolutionary arguments when he discusses the developmental systems perspective (Gray 1992). "Developmental systems" with ontogenetic as well as evolutionary importance are the very links between developmental and evolutionary discourse. But this free switching has its drawbacks as evolutionary aspects may come to dominate the discussion almost imperceptibly.

A central problem in recent discussions about "Developmental Systems Theory" is the "boundary" problem, or the problem of how to delineate an individual developmental system. This problem has been introduced as follows: "Elvis Presley is part of my developmental system, being as he was causally relevant to the development of my musical sensibilities, such as they are. Yet surely there is no system, no sequence, no biologically meaningful unit, that includes me and Elvis" (Sterelny, personal communication in Griffiths and Gray 1994). So how are the lines to be drawn? Griffiths and Gray tackle the problem by proposing an evolutionary delineation of developmental systems: the interactions that produce outcomes with evolutionary interpretations are part of the developmental system. Thus, blink reflexes have an evolutionary history, but a scar on your hand probably has not; therefore, "the resources that produce the blink reflex are part of the developmental system. Some of those that produce the scar are not" (Griffiths and Gray 1994).

According to this proposal, boundaries of developmental systems depend not on causal relevance but on evolutionary considerations. Consequently, discussions about developmental systems turn into discussions about evolution. The response to Griffiths and Gray by Sterelny, Smith and Dickison (1996) further confirms this tendency. They see developmental systems theory as a characterization of evolution—one of the four existing ways.[6] Like Griffiths and Gray, they think that looking at causal relevance will not solve the boundary problem: no measure of causal relevance will distinguish "the role of Elvis from the role of early nutrition." Evolutionary considerations must do the work, and according to them this leads to a privileged role for the genome. They agree that the genome plays no unique causal role in development. It is privileged not because of its causal role in development but because of its adaptive history: "Only some elements of the developmental matrix are adapted for their role in development." Since evolutionary relevance cannot be assessed on ontogenetic causal grounds, the definition of developmental systems is thus decoupled from the causes of development. Developmental systems theory, developed in this direction, becomes increasingly focused on evolution.

The boundary problem involves the realistic insight that not everything can be included in explanations, and that decisions have to be made. This is nothing special, it is true of every explanation. As argued in Chapter 3, a complete causal explanation is never possible. But there is not one and only way to make the choices. If you are interested in evolutionary aspects of developmental systems, you can make the choices the preceding authors suggest. Evolutionary concerns naturally lead to emphasis on elements that have continuity across the generations. For other purposes, the lines can be drawn differently. Development happens once for every organism.

Everything that is influential during that unique period is a causal factor of development. This does not imply that the causes of every old scar on a leg are an interesting object of study. The point is the more general one that evolutionary continuity cannot be the criterion for causal importance. A toxin that accidentally kills an embryo need not be evolutionarily interesting, but it is a devastating causal influence in the development of that embryo.

All in all, from the point of view that environmental influences in development deserve attention, the integration of development and evolution is not automatically a blessing. The neo-Darwinian role for the environment is to select among organisms; this association discourages consideration of a direct environmental role in development. Constructionists reject the causal privileging of internal factors in development and propose to take in environmental influences. But when, in constructionist views of integration, evolutionary concerns become dominant, developmental causation tends to disappear as a subject in its own right. Therefore, though development and evolution are relevant for each other, the goal should not be to blur every distinction. Causal questions about development should not be answered on the basis of evolutionary notions and questions.

5 ✒ Environmental Causes in Ontogeny

This chapter continues the argument that causal choices are unavoidable. Specifically, conceptual tools aimed at the study of environmental influences in development fit in nicely with a constructionist approach. The second part of the chapter provides examples of environmental influences in ontogeny.

Conceptual Tools: Norms of Reaction

In Chapter 3, I argued that neither explanations nor theories can be complete; they always have their restrictions. Restrictions contain a danger: that the choices involved are forgotten and that some dominant factors of study come to be viewed as the whole causal picture, or its most important part. According to Levins and Lewontin, this is what has happened in the case of the Cartesian reductionist world view. In *The Dialectical Biologist* they argue that Cartesian reductionism, which sees any whole as caused by its parts, is not only a method, but a way of looking at the world. This amounts to a confusion of a method with an ontological view (Levins and Lewontin 1985, 2; see also Lewontin 1983a).

But while Cartesian reductionism makes a particular kind of reduction ontologically special, it should not be concluded that asking restricted questions as such is a vice. A scientific question necessarily involves restricting choices concerning the complexities of the world. It is important to be aware of such choices, and to beware of thinking that some favorite choices are the only ones possible, but it is impossible to do without restrictions; you cannot pay attention to everything at once or in all possible ways at once. I propose in this chapter to look specifically at roles that the external environment can play in development and I will argue that this fits in with a constructionist approach.

Questions concerning the influence of the environment clearly have their limitations: they are not asking for the complete process. They are not reductionist, however, in the Cartesian sense indicated by Levins and Lewontin: they are not asking for an analysis of an organism in terms of its parts. Also, I do not intend to give them ontological priority by saying that they are the most important questions about development; the view that environmental influences should be studied does not imply that genes are unimportant or can be neglected.

"Norm of reaction" is a relational concept in the study of environmental influences in development. It refers to the phenotypic outcomes of a certain genotype in a range of external environments. Use of the concept is advocated by some critics of genetic program approaches, such as Levins, Lewontin, and Gottlieb. Lewontin, especially, has been advocating use of the concept in many of his writings. The primary context of discussion is not development, but evolution, and this makes a difference in some respects. But let me first give a general introduction.

"Norm of reaction" is not a particularly new concept; it was coined by Woltereck in 1909 to describe cyclical changes in *Daphnia* related to environmental factors, probably a chemical released by a predator. Since then, the concept has lived a life in the shadow of more dominant concerns. In a historical overview, Stearns noted that traditional evolutionary biology has not focused on the processes generating phenotypic variation (Stearns 1989) and that "until recently, the place given to reaction norms in evolutionary thought was mostly ceremonial" (p. 439). The time has come to change this, according to him.

Stearns mentions David Suzuki et al.'s *Introduction to Genetic Analysis* (1986; later edition Griffiths et al. 1993) as a favorable exception, because it does pay attention to norms of reaction. In the introductory chapter of this book, of which Richard Lewontin is one of the authors, it is stated that "a general view of the interaction of DNA with the environment is a necessary prelude to the detailed analyses that are found in the chapters ahead" (Suzuki et al. 1986, 3). The book addresses the question how the relation between genotype, environment, and phenotype can be quantified. The answer is, through reaction norms. For a particular genotype, a table could be made showing the phenotype that would result from that genotype in each possible environment, and this tabulation is the norm of reaction of the genotype.

The authors note that in practice only partial tabulations can be made: for a partial genotype, a partial phenotype, and some particular aspects of the environment. For example, eye size of a fruit fly could be specified as a function of various constant temperatures for several different partial genotypes (p. 6). They also stress that the phenomenal successes of genetics are due to analyses of genotypes of which the norms of reaction do not overlap. The variation caused by the environment is not a source of confusion in such cases; consequently, the role of the environment can virtually be ignored, because each genotype yields a distinct phenotype irrespective of the environment. However, such cases are not the most common ones, the authors say. For example, among the hundreds of classical mutants known in *Drosophila*, only a quarter ideally suit the purpose of genetic analysis. Though the conceptual world of experimental genetics is dominated by such one-to-one relationships between genotype and phenotype, they do not represent the most common situation in the natural world (Suzuki et al. 1986, 10).

The chapter on quantitative genetics stresses the importance of reaction norms in more detail, defining the concept as follows: "The way in which the environmental distribution is transformed into the phenotypic distribution is determined by the norm of reaction." A norm of reaction is like a distorting mirror that reflects the environmental distribution onto the phenotypic axis (Suzuki et al. 1986, 514). When organisms are phenotypically plastic in a certain characteristic, that is, when different phenotypic outcomes result in different environments, the norm of reaction for this characteristic has a nonzero slope. See Figure 5.1.

The authors state that remarkably little is known about the norms of reaction for any quantitative trait in any species, and that

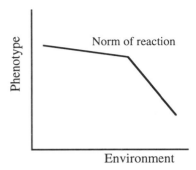

Figure 5.1
Norm of reaction (of arbitrary shape)

no norm of reaction is known for any quantitative trait in humans. From the little that is known from species such as the fruit fly and maize it is clear, however, that no genotype's reaction norm is consistently above or below that of other genotypes. Consequently, selection of "superior" genotypes will result in varieties that are specifically adapted to the environment in which they have been selected for; they are probably not superior in all other environments. Further, when there is genetic variation for traits, it is unlikely that one genotype is favored over another across a range of environments. Applying these insights to judgments on humans, Suzuki et al. conclude that "there is simply no basis for describing different human genotypes as 'better' or 'worse' on any scale, unless the investigator is able to make a very exact specification of environment" (p. 518). See Figure 5.2.

Lewontin has stressed that a consideration of reaction norms should lead to caution in the interpretation of analyses of variance (ANOVA). When variance is analyzed into components assigned to genotype, environment and genotype-environment interaction, the results depend on phenotypic population means and are therefore local. In Figure 5.2, analysis of variance in the environment on the right will yield significant genotype effects, while in the environment on the left significant interaction will probably be the result.

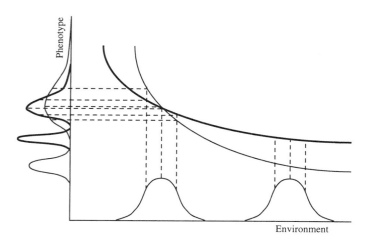

Figure 5.2
Two different genotypes with crossing reaction norms, represented by a thick and a thin line. The environmental distribution on the right (horizontal axis) results in two discrete phenotypes while the left environmental distribution yields phenotypic distributions which cannot be distinguished (after *An Introduction to Genetic Analysis*, 3rd ed. by Suzuki, Griffiths, Miller, and Lewontin © 1986 by W. H. Freeman and Company. Used with permission.)

Extrapolation of such analyses to the effects of new environments cannot be done without knowledge of reaction norms. The specific reason for caution is that it cannot be assumed that reaction norms of different genotypes are globally parallel (Lewontin 1974b; Gupta and Lewontin 1982).

The discussion about the partitioning of variance and the problem of crossing reaction norms takes place in an evolutionary context, where the heritability of characters is a main topic. If reaction norms cross, ANOVAs can only be interpreted locally, and conclusions about heritability are restricted to that environment or else worthless. Given that little is known in most cases about the shapes of reaction norms, there is plenty of room for discussion about their need or usefulness. If one assumes that reaction norms usually cross in normal environmental ranges, extreme caution in the interpretation of ANOVAs is needed. If crossing is rare, however, a heavy emphasis on reaction norms seems unnecessary. In a defense of the use of heritability estimates, Neven Sesardic writes that "the more uncertain the frequency of G-E [gene-environment] interactions, the more dubious is the alleged usefulness of the norm-of-reaction approach" (Sesardic 1993, 407).

Whatever may be true concerning the crossing of reaction norms, it has only limited relevance for developmental biology since not heritability but development is the primary issue there. Concerning development, attention can be restricted to one genotype at a time. Independent of their role in discussions of heritability, reaction norms are extremely useful to depict environmental influences in ontogeny.

A norm of reaction relates phenotypic outcomes to a range of environments for a given genotype, so it describes nothing like the complete process of development. Russell Gray has criticized the concept because of this restrictedness. He argues that though norms of reaction are an improvement over older views, they are still too static and one-sided. They do not acknowledge the changing influences of an environment in the course of development, nor changes in the environment itself. A better picture of development, says Gray, would "include a range of life-history trajectories over a range of codeveloping environments" (Gray 1992, 174).

Gray's intention is to develop a view that does more justice to the relational complexities of (evolution and) development. His rejection of reaction norms has to do, I think, with his calling the concept a "view" (see previous paragraph). I agree that as a view of development it would certainly be too limited. But although research tools and views are often intimately related, and although there is a danger that limited research tools are "ontologized" when they become too dominant, a complete identification of views and tools cannot be maintained. Within an overall view of some complex process one is often forced to reduce complexity, conceptually and/ or experimentally, in order to answer specific questions. The usefulness of available tools depends on the questions that need answering.

Specifically, within the view that interactions are the important causal elements of development, there is a need for practical conceptual research tools. I see the study of reaction norms as very useful in this era of genetic dominance. As things are now, I do not think there is even a slight danger that the tool will become a view that would make people erroneously think that embryological development is uniquely caused by the environment.

Often, norms of reaction are distinguished from other types of phenotypic plasticity, to which I now turn.

. . . And Polyphenisms

"Phenotypic plasticity" is a general description of environmentally induced phenotypic variation. Reaction norms, in their broadest definition, cover all phenotypic plasticity. The concept is sometimes defined in this general way, as the set of all phenotypes that are displayed by a single genotype under different environmental conditions (see Van Noordwijk and Gebhardt 1987, 75). Following this definition, reaction norms may be of any form. This includes flatness: flat reaction norms, which represent invariable outcomes under different environmental circumstances, are explicitly described as reaction norms sometimes (Sultan 1987, 131).

The general definition of reaction norms is not the only possible one. Within the field of phenotypic plasticity, reaction norms are often distinguished from polyphenisms. This distinction can be made in two ways. First, reaction norms may refer to cases where continuous variation in some aspect of the environment yields continuous variation in phenotype. This is how Stearns uses the term: "When an organism produces a phenotype that varies as a continuous function of the environmental signal, the relationship is called a reaction norm" (Stearns 1989, 436). Polyphenisms on the other hand refer to discontinuous functions.

The second way to make the distinction between reaction norms and polyphenisms overlaps with the first one but refers to underlying mechanisms. Nijhout, in his book on butterfly wing patterns (Nijhout 1991), distinguishes three responses to temperature changes in development: norms of reaction, seasonal polyphenisms, and phenocopies (to be discussed later). Norms of reaction describe the direct effects of the environment, particularly of temperature. Since biochemical reactions and physiological processes go at different rates at different temperatures, morphological outcomes are bound to change, too (Nijhout 1991, 121). Here, norms of reaction are defined not by continuous outcomes, although continuity is expected, but through the underlying mechanism. Nijhout does not go further into norms of reactions because too little is known about them in the context of pattern formation on butterfly wings. In the case of seasonal polyphenisms, which Nijhout describes as "adaptive alternative morphs," the environment serves as a stimulus to effect a discrete developmental switch to an alternative morphology. Seasonal polyphenisms are widely seen among butterflies.

Thus, norms of reaction are distinguished from polyphenisms either by a continuity-discreteness distinction or by the mechanism causing those outcome patterns: continuous slowing down and speeding up processes in the case of norms of reactions, and sudden switches into alternative pathways in the case of polyphenisms. The two criteria are often assumed to coincide. In addition, polyphenisms are assumed to be adaptive, while norms of reaction are not necessarily so.

But some care is needed, since in most cases little is known about the mechanisms. There has been much speculation, mainly in relation to questions on how these phenomena evolve. For example, many authors have postulated mapping functions that "translate" morphogenetic substances into phenotypic outcomes in order to explain canalization phenomena.[1] However, in a review of the literature Scharloo (1991) has concluded that the assumption of the existence of such mapping functions cannot be maintained, and that very little is known in fact about the phenomenon of canalization. The mechanisms involved in polyphenisms, norms of reaction and their evolution are also virtually unknown. Shapiro characterized his attempt to work out the genetics of the epigenetic system controlling seasonal polyphenism in a group of butterflies as a "noble failure" (Shapiro 1984). Concerning norms of reaction, too, much speculation exists on possible mechanisms from an evolutionary and quantitative genetics point of view. Adaptiveness cannot be assumed, as Scharloo has emphasized: "designating a reaction norm as adaptive is only justified when you know how it works" (Scharloo 1989). In many cases they may simply result from physiological or other constraints.

Some researchers assume that special genes for plasticity must exist, in addition to the genes that influence the plastic traits, while others question this (Via 1987, 1993; Scheiner and Lyman 1989; Schlichting and Pigliucci 1993; Scheiner 1993). As Sultan (1987, 171) has argued, a better understanding of developmental mechanisms, including mechanisms of gene expression, is needed for a successful integration of plasticity into evolutionary theory.

A further qualification of the distinction between discrete polyphenisms and continuous reaction norms comes from the fact that discrete phenotypes may result from continuous mechanisms. When phenotype is a continuous function of temperature, and a species has two rounds of reproduction in a year at temperatures that are wide apart, two discrete phenotypes will be the outcome. This is the case, for example, in butterflies of the genus *Bicyclus* (see Brakefield and Reitsma 1991).

However, it is not my intention to suggest that the distinction between polyphenisms and reaction norms is worthless, only that it cannot be taken as absolute. Continuous and discrete forms of plasticity, or norms of reaction and seasonal polyphenisms, both represent environmental influence in ontogeny and deserve further study.

Epigenetics: Study of Gene Regulation

Epigenetics, the study of gene regulation, is a field in which such further study could be carried out. It is true that epigenetics is often gene-centred. For example, in an overview in "Developmental Genetics," Robin Holliday says that epigenetics has as a key feature the unfolding of the genetic program. Nevertheless, even the most gene-centered interpretation of epigenetics goes beyond pure genetics. Holliday rejects the view of the "optimistic" school of developmental geneticists, which holds that the accumulation of information on genes will automatically reveal the whole mechanism of development. The alternative is that epigenetics should be a real discipline, which, he adds, is not necessarily a pessimistic view (Holliday 1994). Rollo (1994), too, distinguishes an epigenetic from a purely genetic approach. The latter, he says, is inconsistent with many findings about regulation phenomena.

Regulatory processes in eukaryotes include DNA-methylation and other aspects of chromatin structure, transcription factors, hormones, mRNA-processing, and DNA-rearrangement by way of duplications, transversions, and transposable elements. Let me say a little about chromatin. Chromatin is a complex of DNA intermeshed with proteins. It can be in different states, and one factor in this is thought to be the methylation of DNA bases, particularly cytosine. When chromatin is in a highly condensed form, called heterochromatin, DNA cannot be transcribed. Euchromatin is a less condensed state in which the DNA is available for transcription. The genome in a cell has certain parts activated, others repressed, and yet others in in-between states. Holliday (1994) calls a cell's particular state of gene regulation its epigenotype. In the same spirit, Jablonka and Lamb (1989) call the chromatin structure of a gene the gene's phenotype. The epigenetic structure is heritable in cell lineages; therefore, heritable defects in this structure are possible of which the effects resemble the effects of mutations (Holliday 1987). This has consequences for evolution, as noted in Chapter 4, but even more directly for development. Holliday notes that if normal DNA methylation is essential for the normal control of gene expression during development, defects in the methylation structure could have severe phenotypic consequences and are likely to be the cause of teratogenic abnormalities.

That chromatin structure is involved in gene expression has been known for some time, particularly through the phenomenon of imprinting. In imprinting, the epigenetic structure of the gamete from the mother is different from that of the father, and the result may be that their genes are differentially expressed. In other words, in imprinting it matters whether a certain gene comes from the father or the mother. This phenomenon has been reported for an ever increasing number of genes and diseases (Peterson and Sapienza 1993; see also De Pomerai 1990; Gilbert 1994). It is a special case of the general phenomenon of changes in a gene's or genome's epigenotype or phenotype.

Environmental influences are not necessarily included in the study of gene regulation (see for example Moehrle and Paro 1994; Bestor et al. 1994). In search of the causes of changes in chromatin structure, for example, the focus is often on genes that control it; the 'imprinting genes' in cases of imprinting (Peterson and Sapienza 1993). But as chromatin structure can also be influenced by external causes (Holliday 1987, 1990, 1994; Jablonka and Lamb 1989; Jablonka et al. 1992), such influences also deserve study. For example, folic acid is possibly involved in supplying methyl groups for DNA methylation. Shortage of folic acid during development is a common cause of neural birth defects in humans. Dietary supplements of folic acid are therefore recommended to women who want to become pregnant (Douglas 1993).

More generally, environmental influences are important causes of birth defects (Braun 1987; Lie et al. 1994), but much is unknown about them. Lie et al. write that although genetic explanations of birth defects are dominant, their study finds strong, if indirect, evidence that environmental factors contribute to the familial risk of birth defects, "suggesting that important environmental teratogens have yet to be discovered" (Lie et al. 1994, 4).

Heat-Shock Proteins

In 1935, Goldschmidt described the production of developmental abnormalities in *Drosophila* following abnormally high temperatures during development. The abnormalities were similar to the effects of specific mutations and he therefore called them "phenocopies."[2] In the 1960s, it was found that heat shock in *Drosophila* development is followed by chromosome puffings (indicating gene activity) at specific sites. Later, this activity was shown to result in the production of proteins not normally found in the cells (see Bournias-Vardiabasis and Buzin 1987). Thus, apart from interfering with normal gene expression, heat-shock also induces the production of special proteins, called heat-shock proteins. The heat-shock response is

found in a wide variety of organisms. Research into the phenomenon is now a major way in which environmental regulation of gene
expression is studied.

Heat-shock responses are found both in adult organisms and
during development. In development, their presence has been impli-
cated in different ways. First, several heat-shock proteins are made
during normal development, in the absence of heat-shock, and are
regulated, for example, by the developmental hormone ecdysone
(Bond and Schlesinger 1987). Second, the response to heat-shock
varies with different stages of development. Specifically, in many
organisms the response is not found before the blastula stage (Bond
and Schlesinger 1987; Kim et al. 1987; Walsh et al. 1993; Heikkila
1993a, b; Infante et al. 1993). The response to mild heat-shock is
known to protect the cell in a certain degree against further damage
from heat (Walsh et al. 1993; Petersen 1990). On the whole, however,
much is still unknown about the mechanisms and functions of heat-
shock responses.

Production of several heat-shock proteins is also found after
other kinds of stress, such as mechanical stress, heavy metals or other
chemicals, which often cause damage in similar ways as heat (Petersen
1990; Atkinson et al. 1983), though the precise set of induced proteins varies with the character of the stress. Therefore, the more
general label of stress-proteins is sometimes used for heat-shock
proteins. Heat-shock is sometimes considered to be a good model
for teratogenesis.

Temperature-dependent gene expression often plays a role during normal development, too, for example at the beginning and end
of diapause. In the silkworm *Bombyx mori*, the end of diapause requires
an enzyme that helps to convert sorbitol to glycogen. The gene for the
enzyme is cold-inducible: after two to three months at 5°C, the gene
is activated (Niimi et al. 1993). There is much uncertainty as to the
precise mechanisms. In *E. coli,* ribosomes are probably involved as
sensors of the signal linking the environmental stimulus with expression of heat-shock genes (VanBogelen and Neidhardt 1990).

Let me take stock. In Chapter 3, I distinguished two ways of
doing justice to relations in development. One is to take differences
as the explanandum, and look for the difference that makes the

difference. The study of norms of reaction is a tool for this; it yields insight into environmental differences that make a phenotypic difference. A second approach to relational causation is to concentrate on mechanisms involved in the process; here, study of the mechanisms of environmentally induced gene expression fits in. The two strategies are compatible; for example, one could first establish some norm of reaction, and next study causal steps in the process. Both strategies involve choices; none of them can possibly cover the developmental process as a whole.

In Chapter 1 I argued that environmental influences need not only general acknowledgment but also specific attention. The second part of this chapter is meant to contribute to this goal with further examples of environmental influences in animal development, in addition to the earlier examples given in Chapter 1 and this chapter. The examples can be found in diverse places in the biological literature such as journals on zoology, ecology, natural history, and evolution. Chapter 1 gave three different types of examples: 1) the environment makes a difference for phenotypic (normal) outcomes (*Bonellia*, crocodiles); 2) the environment makes a difference by being a source of disturbance (sexual development in gulls and terns); 3) development of organisms requires a normal environment, often of a very specific character (*Mantispa uhleri*). Examples of each type will follow. Insight into the mechanisms of environmental influences varies, but is often still very limited.

Making a Difference

The environment often makes a critical difference for outcomes of development, either continuously or discontinuously. In some cases partial insight exists in the mechanisms involved. In the following examples I will refer to such insight in several cases. The aim is not, however, to review all the available data. Besides, much is unknown.

• Aphid development. In aphids, great differences exist between the generations through one season. Beck (1980, 102–

105) gives the following description of typical aphid generations. Details vary from species to species, and the seasonal
cycle may also be reduced in comparison with the following
generalized account. See Figure 5.3.

First, there are the eggs, which spend the winter in diapause on the primary host plant. In spring they develop to
become the first generation of females: the wingless fundatrices,
or stem mothers. They give rise to further generations of wingless females, called fundatrigeniae. Stem mothers as well as
fundatrigeniae remain on the primary host plant and reproduce
by viviparous parthenogenesis; that is, live animals are born that
have developed from unfertilized eggs. After a few generations,
females with wings appear, who leave the primary host plant
and settle on the secondary host plant, or (in some species)
other plants of the same species or (in still other species) various other acceptable plants. They also reproduce parthenogenetically and give rise to further generations of parthenogenetically reproducing females, called alienicola, which may be
either winged or wingless. In autumn there appear sexuparae:

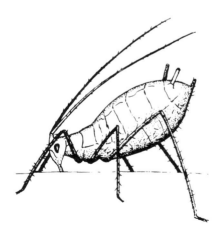

Figure 5.3
An aphid

winged females who return to the primary host plant. They reproduce parthenogenetically, giving rise to the sexual males and females of the species. In most species individual sexuparae produce offspring of both sexes. The sexual females, called oviparae, are wingless; they mate with the winged males, and eggs are deposited on the primary host plant, where they pass the winter in diapause.

Environmental influences determine the appearance of the different morphs. Again, details vary from species to species. The diagram represents a hypothetical series of choices in one aphid, *Megoura viciae*, which Beck derives from the work of Lees (Beck 1980, 114). See Figure 5.4.

According to this hypothetical diagram, the three aspects of aphid phenotype that vary during the season are determined at different points of development. Early in the developmental process, temperature makes the difference between male and female. In female development, further decisions follow, also induced by environmental factors. First, photoperiod and temperature determine whether the female will be sexual (and lay

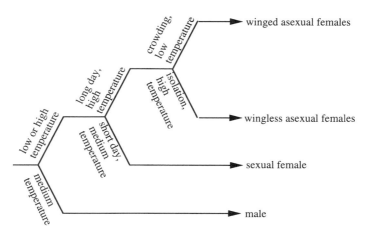

Figure 5.4
Developmental alternatives in the aphid *Megoura viciae* (after Beck 1980)

eggs) or asexual (and reproduce parthenogenetically). Finally, temperature and crowding conditions make the difference between wingless and winged forms.

• Temperature-dependent sex determination in reptiles (see Chapter 1) is studied relatively extensively. The different patterns are illustrated in Figure 5.5.

There has been much speculation about the molecular and physiological mechanisms. A review of hypotheses is given by Janzen and Paukstis (1991). They discuss three hypotheses that all involve the existence of stretches of DNA controlled by a temperature-sensitive effector molecule. In addition, there may be other elements in the process of sex-determination that are influenced by temperature.

• *Mermithids* are a family of roundworms who are free living as adults, but parasitic on insects in the larval stage. In this larval period, they feed on the host, which is killed when the worms emerge from it. The sex of the worms is determined by the amount of food they get, as can be inferred from:

a. Crowding effects. When there are very few worms per host (one to five), they become almost always female. The percentage of males increases with the number of worms per

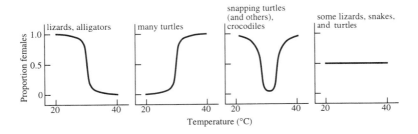

Figure 5.5
Patterns of temperature-dependent sex determination in reptiles (modified after *Evolution of Sex Determining Mechanisms* by James Bull, p. 117 © 1983 by Benjamin/Cummings Publishing Company. Used with permission.)

insect; when there are more than sixteen worms per host, nearly all of them become male. Several studies have demonstrated that the decisive factor is the amount of host available to the developing worm.

b. Host nutrition. The sex ratio is much more female-biased when the host is well-fed than when it is starved (Bull 1983, 111–115).

• Individuals of the snail *Crepidula fornicata* pile up to form a mound (see Figure 5.6). Young individuals, on top of the mound, are always male. After a period of lability, they can become male or female, depending on the underlying individual: if that is female, it becomes male again. When the number of males grows, some of them will become female (Gilbert 1994, 784; see also Gould 1985, 56–63).

• The moth *Nemoria arizona* has two generations a year, one in early spring and one in summer. Larvae have a different diet, depending on the season, and their phenotype differs with the diet. In spring they eat oak catkins, and get the color of these catkins. Summer caterpillars eat oak leaves and become colored like oak twigs. The factor that makes the difference appears to be tannin; catkins contain little tannin, leaves have more of it. If tannin is added to the catkins artificially, caterpillers

Figure 5.6
Cluster of *Crepidula* snails

who eat them get the summer color (Greene 1989; Gilbert 1994, 744).

• Many butterflies show seasonal polyphenisms (see for examples Shapiro 1976; Beck 1980; Nijhout 1991). In temperate regions, where daylength varies greatly, photoperiod often makes the difference between a spring and a summer morph. The European map *Araschnia levana*, mentioned in Chapter 1, is an example. In climates with a wet and a dry season, the wet and the dry season morph can differ markedly. Species of the genus *Bicyclus*, which have been studied in Malawi, have conspicuous eyespots in the wet season, when they are very active. The dry season morph, that is much less active, completely lacks the eyespots and is inconspicuously brown among the dry leaves. The determining factor is temperature at the end of the larval stage. Low temperature yields the dry season morph, high temperature the wet season morph with the eyespots. Intermediate forms result at intermediate temperatures (Brakefield and Reitsma 1991).

• The comma butterfly, *Polygonia c-album*, has a light and a dark morph (see Figure 5.7). Dark adults hibernate and give rise to light as well as dark offspring. Which morph develops

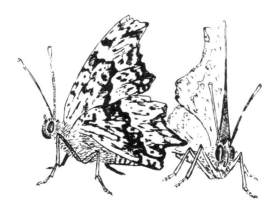

Figure 5.7
The comma butterfly, *Polygonia c-album*; dark and light morph

depends on the photoperiod. Not only daylength matters, but also the change in daylength. Increasing daylength yields the light morph. The light morph is a much quicker reproducer. It gives rise to another generation (of dark butterflies) in the same year, while the dark one goes into diapause in winter, only reproducing the next year (Nylin 1989).

• Phenotypic differences between organisms which originate in the mothers are called maternal effects. These origins can be genetic or environmental. In both cases, hormones produced by the mother are probably main regulators. An environmental example is that photoperiodic influences in the life of the mother often decide over diapause in the offspring. For example, in some populations of the mosquito *Aedes atropalpus*, egg diapause results when during the sensitive fourth instar and pupal stages of the mother there are nine or more short days. Temperature can interfere here: if temperature is high, developmental rate increases and the photosensitive period lasts less than nine days. The result is nondiapausing eggs (Mousseau and Dingle 1991).

• Food is evidently a major environmental variable during development, which may influence the sex of the animal as in the example of the mermithids given earlier. In insects such as bees, female development into two distinct castes is determined by the quantity of food the larvae get (see Brian 1980). In other animals, food may influence growth and maturation rate, maturation size, reproduction or longevity. For example, in the terrestrial slug *Deroceras laeve*, growth rate of juveniles on low quality food is only about a third of the growth rate on high quality food. But poorly fed animals keep growing and eventually become much larger than well-fed ones (Rollo and Shibata 1991).

Food also influences development in cichlid fishes, where differences in diet cause differences in jaw morphology (Meyer 1987).

• The tiny parasitic wasp *Trichogramma semblidis* sometimes lays its eggs in a butterfly host and sometimes in an alder fly host. The outcomes of development in both hosts are strikingly different: an animal with wings in the one (butterfly) case, and an animal without wings in the other (Gottlieb 1992, 153). See Figure 5.8.

The term cyclomorphosis is used by limnologists and plankton specialists to designate seasonal phenotypic changes in algae, rotifers, and crustaceans (Shapiro 1976).

• The water flea *Daphnia pulex* is found in many lakes and ponds, where it eats algae. Young *Daphnias* themselves are eaten by larvae of the phantom midge *Chaoborus*. Juvenile water fleas react to the presence of *Chaoborus* larvae by producing small neck teeth. These teeth, tiny as they are, may protect the animal against predation (Havel and Dodson 1984; Dodson 1989), but it can perhaps not be ruled out that the protection is due to some other characteristic of the induced morph (Spitze 1992). See Figure 5.9.

• Other species of *Daphnia* show continuous variation in the height of their 'helmet' through the year (see Figure 5.10). This phenomenon was observed by Woltereck and, like the previous one, is probably a reaction to a chemical from a predator (Dodson 1989).

• Rotifers or wheel animalcules are extremely small, many of them not being larger than protozoa, but they are in fact relatively complex multicellulars. They have a crown of cilia at the anterior end, which serves for feeding: the rotating cilia

Figure 5.8
The two forms of *Trichogramma semblidis* (after Gottlieb 1992)

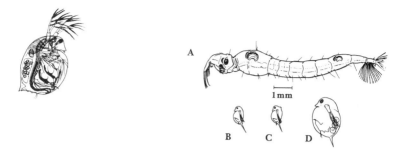

Figure 5.9
Left: a *Daphnia*; magnified compared with the *Daphnias* on the right (drawing Bas Kooijman). Right: the upper animal (A) is a *Chaoborus americanus* larva. B and C are juveniles of *Daphnia pulex*, without (B) and with (C) neckteeth. D is an adult *Daphnia pulex* (right pictures from Dodson, *BioScience 39*, p. 448 © 1989 American Institute of Biological Sciences. Used with permission.)

draw a current of water into the mouth. Rotifers are abundant in fresh water.

Most rotifers reproduce by diploid female parthenogenesis most of the time, but once in a while sexuality occurs, and this is determined by environmental influences. Sexuality starts with

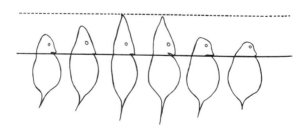

Figure 5.10
Schematic picture of helmet variation in *Daphnia* (from Dodson, *BioScience 39*, p. 448 © 1989 American Institute of Biological Sciences. Used with permission.)

some amictic (female-producing) females producing offspring of a different kind (mictic females), which make haploid eggs. These eggs develop into males when unfertilized, and into encysted dormant embryos called resting eggs when fertilized. The resting eggs become next year's parthenogenetic females.

The exact mechanism of sexuality induction is species-specific. Sometimes photoperiod is the determining factor, sometimes food, sometimes a high population density. In three species of the rotifer *Asplanchna* the amount of tocopherol (vitamin E) in the diet is the crucial factor. In some of these cases, it also matters what kind of prey afforded the tocopherol. In *Asplanchna intermedia*, three female morphotypes can be distinguished; one is characterized by low tocopherol content, one by tocopherol from congeneric prey, that is, other *Asplanchnas*, and one by tocopherol from other kinds of prey (Gilbert 1980). Only one of these morphs (the last one) can be mictic.

• The presence of *Asplanchna* influences the development of another rotifer, its prey species *Brachionus calyciflorus*. *Asplanchna* releases a chemical into the water which causes *Brachionus* to develop a pair of long spines, which it does not possess in the absence of *Asplanchna*. The spines protect *Brachionus* from predation by *Asplanchna* (Gilbert 1966). See Figure 5.11.

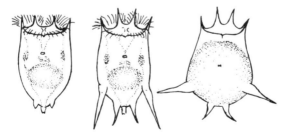

Figure 5.11
Brachionus calyciflorus. Left: a young individual with three short pairs of spines (two anterior pairs and one posterior); no postolateral spines are present. The middle and right figures show individuals with long, *Asplanchna*-induced, postolateral spines (drawn with permission after Gilbert 1966, Rotifer ecology and embryological induction, *Science 151*, p. 1235 © 1966 American Association for the Advancement of Science)

• Apart from effects of predator-released chemicals on morphology, there is also increasing evidence for induced changes in life-history characteristics. In the presence of chemicals released by fishes, two *Daphnia* clones showed decreased size at maturity (first reproduction). When cues from invertebrate predators were present, on the other hand, size at first reproduction increased (Weider and Pijanowski 1993).

• Life histories of the freshwater snail *Physella virgata* are also phenotypically plastic. In the presence of a particular crayfish, they show little reproduction until they reach a size of 10 mm after eight months. In the absence of the crayfish, they grow until they are 4 mm after three and one-half months and then begin reproduction (Crowl and Covich 1990).

• In American toads (*Bufo Americanus*), the size at metamorphosis is a life-history characteristic that changes in response to food and also in response to an odonate predator (*Anax junius*). The larval toads respond to the predator by a considerable reduction of their activity and this in turn reduces growth rate and also size at metamorphosis (Skelly and Werner 1990). Life history plasticity has also been reported for other toads (Newman 1989, 1992).

• *Ambystoma tigrinum* metamorphoses only in the warmer region of its range; the larval form then gives way to the land-dwelling tiger salamander. In colder ponds, the species is neotenic: it retains its larval form. The hypothalamus cannot produce TSH-releasing factor at lower temperatures. In these conditions, the larval form is the reproducing one (Gilbert 1994, 726).

I have concentrated on animals, but phenotypic plasticity is also widespread in plants and unicellulars; in plants it is almost ubiquitous (see e.g., Sultan 1987). Let me give a few glimpses of environmental influences in unicellulars.

• The amoeboflagellate *Naegleria gruberi* can change its form from that of an amoeba to that of a flagellate. During most of its life cycle, *N. gruberi* is a typical amoeba, feeding on soil bacteria and dividing by fission. However, when bacteria are diluted and hence become rare in the environment, the amoeba develops a streamlined body shape and two long anterior flagella within an hour (Gilbert 1994, 10–11).

• The colonial organism *Volvox carteri* reproduces asexually much of the time, but once a year it becomes sexual. This happens shortly before the pond in which it lives dries up. The zygotes can survive the drought and heat of the summer and the cold of the winter. When rain fills the pond again in spring the zygotes break their dormancy and new generations of *Volvox* develop which again reproduce asexually. Temperature induces the sexual phase in this organism (Gilbert 1994, 18–20).

• *Escherichia coli* reacts to changes in temperature and osmotic pressure by changing the protein composition of the outer membrane. This happens through changes in gene expression. Cues from the environment are transmitted by way of an inner membrane protein which in turn influences a DNA binding protein that regulates gene expression. The system clearly shows an interplay of environmental and genetic control (Forst and Inouye 1988).

Abnormal Environments

Chapter 1 gave the example of disturbed sexual development in gulls and terns caused by pesticides. Information on such effects has long been lacking, because toxicological research focused primarily on effects on survival and reproduction in adult animals. Recently, however, effects on offspring are increasingly emphasized and studied. The book *Chemically-Induced Alterations in Sexual and Functional Development: The Wildlife/Human Connection* (Colborn and Clement 1992; see also Fry and Toone 1981; Stone 1992, 1994; Mestel 1994; Culotta 1995) says in its "consensus statement" that many man-made chemicals which have been released into the environment have the potential to disrupt the endocrine systems of animals, including humans. The effects resemble those of DES (diethylstilbestrol), the synthetic estrogen that for many years was given therapeutically to countless women during pregnancy. Daughters born to mothers who took DES suffer increased rates of vaginal cancer, various genital tract abnormalities, abnormal pregnancies, and some changes in immune response (Colborn and Clement 1992, 2). Problems of this type are caused in the embryonic stage but become apparent only later in life. Since they are not visible at birth they can easily be missed for a long time; in many (similar) cases, the causal agent of problems may never be identified.

Several pesticides that are abundant in the environment have such hormone-like effects. Among the developmental effects in wildlife populations are demasculinization and feminization of male fish, birds and mammals, and defeminization and masculinization of female fish and birds. These effects are reported in detail in several papers in the book. Pesticides may mimic effects of natural hormones by recognizing their binding sites on receptors. The hormone-receptor-complex binds to specific regulatory regions of DNA, thus influencing gene expression (McLachlan 1992, 107). The consensus statement mentioned earlier recommends, among other things, that more research is needed on such developmental effects.

Background Environments

All development involves exchanges with the environment. In infinitely many ways, environmental conditions that do not make a difference for phenotypic outcomes are vital cues in the process of development. Here are some rather arbitrary examples.

* Developing American oysters need a hard substrate to land on (Gilbert 1994, 742).

* Some sea urchins settle only on substrate that is covered by a bacterial film; the bacteria produce a factor that induces the sea urchins to settle (Gilbert 1994, 742).

* The adult sand dollar *Dendraster excentricus* (see Figure 5.12) excretes a chemical into the sand that attracts larvae and induces their metamorphosis. The result is that larvae tend to

Figure 5.12
A sand dollar

associate within or near existing sand dollar beds (Highsmith 1982; Gilbert 1994, 742).

• Many mollusks need specific cues for settlement. For example, most sea slug larvae metamorphose only if triggered by live adult prey (Gilbert 1994, 742).

• One subspecies of the spotted skunk mates in April and the young are born two months later, in June or July, when food is abundant. Another subspecies lives at higher altitudes and is not able to mate in April; it mates in September. Implantation of the embryo in this subspecies is delayed until April. The timing is probably determined by daylength. Delayed implantation occurs in numerous animals (Gilbert 1994, 853).

• The maturation and ovulation of mammalian eggs follow one of two basic patterns, which are both dependent on the environment. The first—found in rabbits and minks—is that ovulation is stimulated by the physical act of intercourse. Most mammals follow the other pattern, which is periodic and most notably influenced by the amount and type of light during the day (Gilbert 1994, 817).

• In insects, yolk synthesis in the fat body of the ovary is influenced by juvenile hormone. The secretion of juvenile hormone is probably stimulated by a brain hormone, which in turn responds to environmental influences. In *Drosophila*, the environmental cue appears to be the photoperiod. In the common mosquito, it is the blood meal (Gilbert 1994, 816).

The following examples of the determination of polarity in early development are given in the chapter on mechanisms of development in the fourth edition of Keeton and Gould's (1986) *Biological Science*, but have been deleted in the fifth edition of 1993, which has shifted emphasis more to genetic mechanisms of development.

• The site of sperm entry in the frog egg determines the polarity of the zygote (Keeton and Gould 1986, 822).

• An *Equisetum* spore acquires its polarity—determining the first cleavage, which produces two unequal cells—under the influence of light (Keeton and Gould 1986, 822).

• Under the influence of asymmetric environmental conditions (temperature or illumination), the zygote of the brown alga *Fucus* forms a protuberance on one side, which determines polarity and the pattern of the first cleavage (Keeton and Gould 1986, 822).

In many cases, relations between species are important developmental factors. Here are some examples:

• Certain trees synthesize compounds that closely resemble juvenile hormone. This compound prevents certain insects from metamorphosing. This was discovered after a European bug mysteriously failed to metamorphose on the paper towels of American laboratories. The American trees synthesized the juvenile hormonelike compound (Gilbert 1994, 743).

• Precocene is a compound produced by certain composite herbs that induces premature metamorphosis of certain insect larvae into sterile adults. Precocene causes the selective death of corpora allata cells, which synthesize juvenile hormone, in the immature insects (Bowers et al. 1976; Gilbert 1994, 743).

• The development of *Mantispa uhleri* (see Chapter 1) is also an example in this category, and so is the development of many butterflies. Food requirements are sometimes very specific;

caterpillars of the purple hairstreak, *Quercusia quercus*, feed only on oak leaves, for example. Such environmental factors do not make a specific difference for the phenotype, but do make a difference between life and death.

• A final example of how normal development depends on the environment comes from the study of brain development in humans and other animals. The system of connections in the brain is increasingly found to depend on incoming signals and on use. Carla Shatz (1992; see also Stryker 1994) gives an overview. An important question is how the connections between the neurons, including connections with all the organs, develop with precision. One line of thought is that the brain's entire structure is recorded in the DNA, but research during the past decade has shown that brain development follows different rules, says Shatz. Brains must be stimulated in order to develop.

Research has concentrated on the visual system. Axons grow along their pathways by sensing a variety of surface molecules on the cells along the pathway. But when they have arrived at their target, there are still decisions to be made about where to connect. The initial connection of axons, or address selection, is not precise. Fetal axons develop many side branches, most of which disappear later in development. The proposed mechanism is that the neurons are not specific in their initial connections and that specificity develops by some kind of competition between the neuronal inputs. Research in cats has shown that covering the eyes of newborn cats disturbs the formation of normal connection patterns. This suggests that the pattern forms as a consequence of use. Special synapses strengthen or weaken connections, depending on whether they receive (correlated) inputs or not.

The foregoing applies to life after birth, when the eye is used. But something similar happens in the unborn animal. Here, retina ganglion cells fire spontaneously in a predictable and rhythmic pattern, and this provides the input that is necessary for proper development. When firing is prevented by a

toxin, side branches of the axons do not disappear; no dominance patterns are formed. So even before the eye is really used
for seeing, its activity is required for proper development. Fine-
tuning of the maturing system is therefore accomplished by
experience. The implication is, Shatz remarks, that the exact
specification need not be laid down genetically, which would
require extraordinary numbers of genes. The proposed mechanism is far more economical.

Lickliter and coworkers study sensory aspects of bird development. They found that experimentally induced unusual
sensory stimulation before birth, both auditory and visual,
influences development. Such stimulation results in changes in
the hierarchical relationship between hearing and seeing in the
young bird (this is to say, which of the two the bird depends
on more). They concluded that the cloistered and simplified
prenatal environment in the egg contributes to the development and organization of early perceptual capabilities (Lickliter
and Banker 1994).

Toward an Ecology of Development

Animals live in an environment and depend on it in many
ways. If you are an American oyster and there is no substrate for you
to land on, development ceases and you perish. In terms of the
distinction between causes and conditions, the environment is a mere
condition or "background condition" in this case. It must be there
for development to continue, but does not make a specific developmental difference, such as between having two legs or four, between
becoming brown or green, male, or female. That the normal environment matters at all tends to be disregarded in developmental
biology. For example, when it is said that embryos are protected from
the environment by egg shells or wombs, this is true, but it does not
imply that there is no environment; rather, wombs and eggs afford
very specific, safe, predictable environments. Also, when environmental differences do not make a difference for the outcome, this does
not imply that the environment plays no role. When the outcome of

development is relatively invariant, this only means that the norm of reaction is flat, not that control is genetic (Sultan 1987, 131).

If you are interested in the process of development in a wider than just a genetic sense, the American oyster's substrate is just as well part of the developmental system as its genes. Things would have been different if it had had other genes. Things would also have been different if there had been no hard substrate.

Perhaps no biologist disagrees with this. Yet it is not hard to imagine why the normal environment nevertheless tends to disappear from sight: it is precisely because it is not assumed to make a specific difference, apart from the difference between life and death. Of course, an embryo needs oxygen and water, and in order to protect certain species of butterflies it may be necessary to protect certain species of plants on which their caterpillars feed, but what is there to study? The answer to this question is implicitly present in the foregoing and has different elements. First, if it were completely obvious what kind of normal environment animals need, there would be no (more) reason to study background environments. But in many cases of species decline, the reasons are not clear and they may well involve environmental dependencies, for example on other species. It is certainly true that insight in specific requirements can yield only limited understanding of specific developmental mechanisms. On the other hand, research into the ways in which developing animals react to normal environments and (normal) changes in variables such as food, photoperiod, temperature, or population density has revealed and may continue to reveal specific phenomena and mechanisms.

Second, upon closer inspection of environmental variation, the environment may turn out to make more differences than had been expected. Many of the cases in which phenotypical differences result from environmental differences have come as surprises, and certainly many more surprises are to be expected when more time and money is devoted to environmental aspects of development.

Third, the study of abnormal factors such as pesticides in development may also reveal highly specific mechanisms and effects. Knowledge of such effects is vitally important for responsible health and conservation policies.

Approaching development as a phenomenon that takes place in an environment, and including the environment as a relevant variable, amounts to an ecological perspective. Within such a perspective genes continue to be crucial difference makers, and questions about genes in development have an important place. The aim of an ecological perspective should not be to downplay genes but rather to make conceptual room for other questions as well. As an integrative framework it is an alternative for a genetic program perspective.

6 ✐ Ethics of Attention

In a criticism of "neurogenetic reductionism," Steven Rose notes that such reductionism offers specific answers to the question "where, in a world full of individual pain and social disorder, we should look to explain and to change our conditions" (Rose 1995, 380). Through its neurogenetic explanations, which point to pharmacology and molecular engineering as solutions for pain and disorder, this approach has important social consequences which he opposes, and he rejects neurogenetic reductionism for that reason.

Rose's view illustrates one way of relating science and morality: look at the moral consequences of scientific choices and on this basis recommend particular directions in science. Other existing ways to link science and morality are to be normative in a different way, for example on a religious basis, or to take a descriptive approach, for example in order to find out how moral values or political views codetermine scientific choices. My approach is Rose's: a normative approach based on an evaluation of consequences. Alan Garfinkel has given a general philosophical defense of an evaluation of science along such lines, starting from the necessary incompleteness of explanation and the consequent unavoidability of choices. I will begin by sketching this point of view, and continue with some examples of normative positions in science based on a consideration of consequences. I will argue that it is important that consequences of scientific choices are studied empirically. Finally, I turn to an evaluation of developmental biology. The previous chapter may have shown that environmental influences in development can be studied, but is such study desirable? In view of the consequences of choices in developmental biology, I think it is.

In order to avoid confusion it is perhaps helpful to point out that this chapter thus deals with two fields of empirical study: the first one is developmental biology; the second one is the study of the morally relevant consequences of particular ways of doing developmental biology.

Garfinkel and Value-Laden Explanations

In Chapter 5 of his book *Forms of Explanation*, called "The ethics of explanation," Alan Garfinkel (1981) argues that science cannot give value-free explanations. His point of departure is Weber's view that science is like a map which can tell us how to get somewhere but which cannot tell us where we should go. This view is shared by Hempel and others who hold that a division of labor exists between value-free causal accounts, given by scientists, and value judgments, made by policy makers and others. Garfinkel disagrees. If scientists are concerned with statements of the form "A causes B," he says, this implies by no means that science is value-free. The reason is that causal statements are always made in a social context. It can be wrong to make true causal statements. For example, it is morally wrong to say, "If you look in the attic, you'll find Anne Frank" in the presence of Nazi search parties, even if it is "merely" a causal statement. It is not the presence of value-words but the context that makes the statement nonneutral (Garfinkel 1981, 137).

Simple as it is, says Garfinkel, the point is often missed by scientists who want to forget about the context of their research, as in war industries. When the context of relevance is not so clear, and when applications of research are not in sight, it is more difficult to evaluate the situation; after all, "any fact may end up aiding some evil cause. So what are we to do?" (p. 138). But even in seemingly neutral situations, Garfinkel maintains, "A causes B" is not value-free. Why not? It is clear that values intrude in causal explanations when a value-laden motive exists to favor some causes, and leave other ones in the background. But Garfinkel focuses not primarily on motives for choices but on the choices themselves and their implications.

Ever since Mill, he argues, it has been clear that explanations typically mention only one or two causal factors as the cause of a phenomenon, giving all the other factors the status of background. Garfinkel's approach to explanation is related to Van Fraassen's, in that it concentrates on why-questions which he analyzes with the help of contrast classes that specify the difference to be explained. For the present discussion the important point is that Garfinkel emphasizes the incompleteness of explanations in a way that fits in with Chapter 3. Explanations unavoidably involve choices.

The incompleteness of explanations becomes morally relevant when practical choices must be made. Any explanatory framework recognizes only certain alternatives and therefore guides you to specific solutions. Thus, not the motives of scientists are at issue, but the explanations themselves: "The value ladenness is a fact about the explanation not its proponents. It is value laden insofar as it insists (. . .) that change come from this sector rather than that" (p. 141). The unavoidability of explanatory choices thus makes science inherently nonneutral because of its social consequences. Clearly, Garfinkel's argument depends on the implicit assumption that science cannot be separated from wider contexts.

Nonmoral Motives: Positivity Bias

Garfinkel's view does not associate value ladenness with the service of moral or political goals. Not motives but consequences count. It is therefore unlike some other forms of value ladenness, for example as defined by Robert Proctor in his book *Value-Free Science?* For Proctor, nonneutrality refers to the explicit service of moral and political ends, to science that "takes a stand" (Proctor 1991, 10).

These two forms of value-ladenness do not always coincide. You may pursue neurogenetic explanations for no other reason than that you happen to work in a particular tradition or because of particular moral motives on how to change the world; yet with respect to the consequences of your work this difference in motivation may be quite irrelevant. Let me explain in a little more detail why value-ladenness of the Garfinkel-Rose type does not presuppose that moral motives are pervasive.

It has been argued that science has an unavoidable input of contextual values. For example, Helen Longino's *Science as Social*

Knowledge (1990), discusses how "contextual" values—that is, values that are not internal to science—unavoidably enter science as part of the background assumptions that influence the evaluation of theories. Wim van der Steen (1995) agrees that contextual values are always present, though he argues that Longino has not really shown their inevitability. According to him, extrascientific values are relevant due to methodological trade-offs, which cannot be avoided because theories cannot satisfy all methodological criteria. Science itself cannot provide adequate guidelines for making a trade-off in a particular way. That is to say, there is no science-internal reason to prefer, say, generality to realism. Therefore, extrascientific purposes have a legitimate role in choices and hence science has intrinsic links with contextual values (Van der Steen 1995, 29).

Extrascientific purposes are indeed important in science, but they need not always be consciously present. Scientists are generally working in existing traditions, where the relevant choices have been made long before. Mechanisms that are not necessarily intentionally moral play a role in confirming those choices. An example of such a mechanism, known from cognitive psychology, is the phenomenon of "positivity bias" (see Evans 1989), the tendency to concentrate on what is there (to see or read or hear), rather than on what is missing. People find it difficult to think outside the domain of the known, about "negative information"; they "seem much better at making use of relevant information which is presented to them than at devising appropriate strategies to discover such [new] evidence" (Evans 1989, 63). Falsification, for example, is a relatively difficult goal psychologically because of this tendency to think in terms of what is positively there, in this case the hypothesis at hand.

Positivity bias can explain details of human reasoning, but I think it can also partly explain, more generally, why some issues tend to be studied in ever growing detail, while other ones remain completely out of sight: subjects of research are self-reinforcing. Subjects that are not studied do not easily spring to attention, for the very reason that they are not receiving attention. Positivity bias thus results in something like Kuhn's "normal science." For example, it throws light on why evolution is and remains so dominant in the philosophy of biology. Once a set of interesting problems in this field had been defined, such as the nature of fitness and adaptation and the structure of evolutionary explanation, further work has tended to stay with these problems, respond to earlier contributions, and refine the treatments.

Through positivity bias, even severe criticism of existing research, such as criticism of gene-centrism, may remain within the existing framework of concepts and phenomena. Fred Nijhout (1990) opposes the genetic program metaphor and the idea that genes determine development. Instead, development results from the actions of heterogeneous causal factors. On his alternative view genes are necessary but not sufficient, which gives a "working hypothesis at all levels of investigation on the function of genes" (Nijhout 1990, 441). The formulation shows that genes remain central objects of attention, even though Nijhout's purpose is to oppose a one-sided emphasis on genes.

Not going on as usual, but doing something different requires conscious decisions. Positivity bias is a nonmoral mechanism that is important in keeping attention where it is. This nonmoral mechanism may lead to moral consequences, though. Thus, nonneutrality through consequences not necessarily coincides with nonneutrality through the service of moral or political ends.

Causal Choices and their Context

Garfinkel's view implies that moral evaluations of scientific choices require specification of a context of moral concern. Such a specification may be controversial, and this in turn makes scientific choices controversial. It may therefore constantly be tempting to think or hope that there are objective and absolutely right ways to make causal choices in science. I will illustrate in this section that this hope is idle. Causal selection always depends on what you are asking, and controversion about which questions are the important ones cannot be replaced by a mechanical procedure.

Garfinkel calls his own approach of social reality Marxist in orientation, which makes it clear at once that his definition of context is controversial. From this perspective, he says, explaining social differences by individual differences looks like "blaming the victim." For instance, an explanation of why Jones rather than Smith is unemployed, presupposes the existence of unemployment. This existence of unemployment, however, is a more fundamental fact about society, and more important in understanding the social positions of Smith and Jones than their individual differences. Therefore, unemployment should be approached as a structural phenomenon (Garfinkel 1981, 180). For this reason Garfinkel rejects social Darwinism, which answers questions such as "What makes one individual rather than another come to occupy a social position?" in terms of individual differences, preferentially genetic ones. Garfinkel argues that first, genetic differences do not fully explain individual differences,[1] and besides, individual differences underdetermine social differences; they are never in themselves causes of social positions (p. 117). Suppose, he says, that a special tax would be levied on red-haired people. If we assume that red hair (the difference between red and non-red hair) is a genetic trait, would it follow that the resulting poverty of red-haired people would have a genetic cause?

This example has also been discussed in the philosophy of biology. Before considering how Garfinkel himself deals with it, let us see what Fred Gifford (1990), in his paper "Genetic traits" has to say. Gifford does not feel able to draw a firm conclusion. After having proposed that a trait is genetic when genetic factors make the difference, in a certain population, between individuals with and without the trait (see Chapter 3), he discusses cases like the preceding one. His example is skin color rather than hair color. If skin color is a genetic trait in many populations, and if we also know that people with black skin receive a comparatively poor education, what to say of the causes of their (comparatively low) academic success? Both genes and the educational environment "make a difference" here. Gifford reminds us that Dawkins's answer in the analogous case of sex roles is that the "original" genetic difference is the real cause, which "makes itself felt through" the educational system.[2] Presumably, the genetic differences between individuals somehow cause the educational system to treat them differently. Gifford's own inclination is, first, to stress the causal picture as a whole, with both genes and the environment as important factors. However, this is not enough.

When blame and political action are called for, as in the example, it is clear, he says, that "environmental factors of discrimination are responsible for the decreased academic success in the sense that this is clearly where blame should be placed and where we should attempt to intervene" (Gifford 1990, 338). With an eye on the social consequences, he thus switches to normative language, specifically the concept of responsibility. Indeed, his conclusion is that "we must take seriously the notion that a normative criterion must be appealed to here" (p. 339).

Garfinkel reaches the conclusion that discrimination is the more important cause in a more straightforward way. In his example, without the social discrimination red hair would have no tax consequences; therefore the real cause of the poverty is social discrimination (Garfinkel 1981, 117). Claims of genetic causality generally disregard the importance of structural social practices such as discrimination, and therefore give a false picture of the causalities involved, he says. Heritability measures in particular are misleading indicators; they are generally but wrongly taken to represent genetic heritability. For example, heritability measures for poverty based on analysis of variance will be high when poverty is caused by red hair, but they will also be high when poverty is only correlated with red hair through taxes, because analysis of variance cannot distinguish between correlation and causation.

Thus, for Garfinkel, the real cause is the social one, discrimination, not the individual differences that become relevant given the social structure. From his point of view, Dawkins's idea that genetic differences "make themselves felt through" a biased education system, looks like sheer nonsense right from the start. It is the other way around: the social structure itself augments and creates differences.

Though Garfinkel makes the priority of social causes appear as a matter of fact, his conclusion rests on his normative approach to social reality. His evaluation is not different in that respect from Gifford's, who only invokes normativity at a different moment in his treatment of the subject.

A more formal way to come to the same conclusion, that hair or skin color gain causal power only given certain social practices, is through the notion of "screening off." According to this probabilistic concept, "If A renders B statistically irrelevant with respect to outcome E but not vice versa, then A is a better causal explainer of E than is B" (Brandon 1990, 83). In this situation, the probability of (E, given A and B) equals the probability of (E, given A) but not of (E, given B). Thus, the influence that B has on E depends on the presence or absence of A, but the influence of A does not depend on B's presence. Brandon has used the notion of screening off to argue that phenotypes are more important for selection than genotypes, because the phenotype matters in the selective interactions of an organism irrespective of the genotype that is "behind" it: the phenotype screens off the genotype from the reproductive success of organisms. The notion could be used for the present case in a similar way: if you have poor academic success because of poor education because of discrimination because of skin color, the poor education screens off the skin color as a cause: it causes poor academic success irrespective of skin color. Skin color, on the other hand, does not explain academic success irrespective of education.

The notion of "screening off" has the effect that more "remote" causes disappear from sight and proximate (in the sense of nearby in time) causes are foregrounded. For the cases discussed this yields Garfinkel's conclusion, that skin color or hair color would not lead to the effect without a social system that made them important. But screening off looks more objective; the selection of causes appears not to involve normative choices; it apparently does not involve choices at all, apart from the choice for the right causal criterion. In the context of selection, indeed, the notion is used to argue that the phenotype is objectively and absolutely the best explanation of differential reproduction (Brandon et al. 1994.)[3] However, the absence of choices is only apparent. While it is true that selection most directly acts on the phenotype, the criterion of screening off does the required work (causally favoring the phenotype) only by favoring a special kind of question about fitness, namely, what is the most proximate cause. Once that question has been chosen, the answer may involve no further deliberations.[4] But no a priori reason exists why remote causes should be less interesting scientifically. When it is asked what difference certain genes make for fitness, for example, the criterion is irrelevant.

Thus, the criterion of screening off itself, by favoring a specific type of question, involves a choice. This choice is plausible for some purposes, but it can be very implausible in other cases. For example, when someone fails an exam due to low grades, the low grades screen off all other causes, but are they really the most important cause of the failure?[5] Other factors, such as bad work, or the causes of that, are probably more interesting for someone who wants to explain the failure. An absolute-looking criterion such as screening off cannot replace decisions about what is important or interesting in complex causal processes.

Since the use of screening off depends on choices, it does not show that choices can be avoided about "the causes of x." The three different ways of dealing with problems of causal selection in this section show that choices as well as their moral context can be described in various ways and in various degrees of explicitness. Gifford's location of choice is maximally explicit, and coupled to a moral evaluation of the issue at hand. For Garfinkel, important normative aspects of causal selection are given with his Marxist approach of social reality, and are therefore a little more in the background. Screening off, when presented as the only right way to deal with causes, hides the choices on which it is based.

Contested Scientific Choices

Scientists may participate in dominant ways of doing science with or without awareness of the choices involved. They may also search for new subjects, metaphors, and questions that serve their purposes better than current approaches. Particular scientific choices are often recommended and some of them, such as Gifford's and Garfinkel's in the previous section, are normative in the moral sense. Let me give a few more examples, which have in common that they center on how to deal with individual selves in science.

According to Levins and Lewontin, a bourgeois ideology of society and a bourgeois view of nature reinforce each other (Levins and Lewontin 1985). The bourgeois view is individualistic and Cartesian, explaining individuals in terms of atomic parts. They reject this reductionism and opt for explanations that make phenomena part of a larger framework, where social conditions are at least as important as causes of illness as bacteria.[6] In Lewontin's (1991) foreword to the book *Organism and the Origin of Self* (see Chapter 2), he distances himself from the project of the book by saying that the central importance attached to problems of self and identity in biology reflects the ideology of individualism. Instead, organisms should be understood primarily as parts of larger wholes.

Related discussions on the concept of self and its moral implications are going on in the social sciences and philosophy. For example, in *Psychology's Sanction for Selfishness,* Wallach and Wallach (1983) criticize the emphasis in modern psychology on the self and its needs, which sanctions selfishness. A similar message emerges from *Habits of the Heart,* a sociological study in which Americans are questioned about their lives and what matters most to them (Bellah et al. 1985). Individualism shows up as a central theme, and self-reliance as an important challenge in life. The concept of self, according to the authors, has become ever more detached from social and cultural contexts. This has important consequences for morality. If the self is supposed to choose its own values, morality becomes personal and subjective. Others are involved only in the sense that they should also be left free to choose their own values. In *Rethinking Goodness,* Wallach and Wallach (1990) call the contractual ethics that results from this emphasis on self the "minimalist predicament" in ethics, because community care tends to disappear. Obligations are to the self; people feel obliged to satisfy the self's needs, to make it successful and happy. Other selves should not be interfered with in similar strivings, but that is all that remains of care for each other.

Discussions about the place of organisms in biology and of individual selves in the social sciences and philosophy are not identical, but they certainly have points of similarity, not least in their moral and political overtones. So after this short excursion into social science and philosophy, back to biology again:

In his discussion of Leo Buss's (1987) *The Evolution of Individuality,* Scott Gilbert (1992) welcomes Buss's way of looking at the relation between an organism and its cells in an evolutionary way, seeing present-day cell lineages as survivors of selection processes.[7] But he deplores the way in which selection is described, which is in terms of competition. According to Buss, competition between cell lineages has resulted in today's embryos, which represent a compromise between the interests of the whole embryo and of selfish cell lineages. What we see here, says Gilbert, is the world view of Adam Smith and other economic theorists (accepted by Darwin and other biologists), which assumes that each individual is an atom of greedy self-interest. In his eyes, "Buss's cell lineages seem like so many Americans in search of community in this era that John Gardner has called 'the war of the parts against the whole'" (Gilbert 1992, 481). There are other definitions of self and society than those that presently predominate in the West, says Gilbert, referring to descriptions of how in many Asian cultures the value of the individual depends on relationship and interpersonal harmony.

Gilbert argues for a view of evolution which sees the units of selection as sociocentric instead of egocentric, and states that biochemical evidence for this can be found in the evolution of cell-cell interactions, beginning with fertilization among protists (p. 482). The cells that gave and give rise to multicellular organisms are cooperators as well as selfish replicators; there has been selection for cooperation. In doing justice to images of cooperation, says Gilbert, we can account for evolution without economic models in which cooperation has to arise from competition.

Morally inspired proposals about how to approach the relation between self and environment are not always explicitly based on the moral consequences of scientific choices. The moral *backgrounds* of contested choices may be a more explicit concern. However, the fact that choices have a moral background is not in itself a strong argument for change. I think that attacks on moral sources of origin are often implicitly directed at the consequences of the contested choices. Levins and Lewontin as well as Wallach and Wallach are explicit about their concern for consequences; it seems to me that in Gilbert's paper this concern is implicitly present as well.

Ethics of Attention

Both Goodwin and Oyama have argued that a gene-centric biology has adverse social consequences. Goodwin discusses geneticism as part of Darwinism, which sees the living world too much in terms of competition and selfishness. Like Gilbert, he deplores these metaphors and argues that there is as much cooperation in nature as there is competition. According to him, the Darwinian view is incomplete and limited, and based on an inadequate view of organisms. The limitations have contributed, he says, to "the crisis of environmental deterioration, pollution, decreasing standards of health and quality of life, and loss of communal values" (Goodwin 1994a, xiv). Oyama, too, suggests that a better view of the complexities of causation in biology may change the world for the better. Talking about the nature–nurture controversy in psychology, she agrees with statements to the effect that those who are in the position to define the self, control the world. So if biology is used in defining the self, it should be "a biology ample enough to include our whole selves," in the sense that this biology should also talk about the world in which we live (Oyama 1993). Goodwin's and Oyama's positions, familiar by now, are that biology should offer more complete causal pictures by emphasizing organisms (Goodwin) and organisms-in-their-environment (Oyama), respectively.

My own approach has been to start from the conspicuous omission stated in the questions asked in Chapter 1: Could and should something be done about the virtual absence of environmental influences from developmental biology? The question whether something should be done about it from a moral point of view can only be answered, I think, by looking at the consequences, real and potential, of different ways of doing developmental biology. If the three approaches to development did not make a difference for practical purposes, it would not matter much whether environmental influences were recognized or not. But on the assumption that science has practical consequences and that the value-ladenness of explanations is located in the consequences of their being incomplete, the differences between the approaches matter, at least potentially.

Since the consequences of science are not generally thought to be under the influence and responsibility of scientists, arguments that take account of those consequences in science have an uncertain status. It is controversial whether they should be part of science, whether they should be considered only when it comes to "applications," or whether scientists should perhaps stay far away from them altogether. As I see it, the question whether or not such arguments are seen as part of science proper is not the really important one. What is important is that they are addressed and that they have an impact on science. I agree with Robert Proctor (1991) who argues in favor of a political philosophy of science; in his view, epistemological questions (such as, How do we know?) divert attention from questions such as: Why do we know this and not that? Why are our interests here and not there? Who gains from knowledge of this and not that? What is to be done—or undone? These latter questions, he says, are not only political but also more fundamental (Proctor 1991, 10).

I propose that what is needed is an "Ethics of Attention," which concerns itself with the social consequences of choices in science. An ethics of attention does not look at knowledge just for knowledge's sake; it sees science as part of the world. It looks for insight in the social consequences of the allocation and the character of scientific attention. In order to avoid that discussion about these issues becomes too general, too vague, or too moralistic, data about such consequences are needed; general impressions are not enough. Moral issues and considerations are now often suppressed, even by critics of "neutral science," partly because they are controversial and are considered not to belong to science proper, but no doubt partly also because of a lack of reliable data. Consequently, such issues are considered to be too "soft." To remedy this situation, systematic insight is needed.

Relatively much has been written on the social consequences of an overemphasis on genes. The history and future of eugenics is a much discussed issue, as is the present direction of medicine, including psychiatry. Let me sketch some of the issues.

While emphasis on genetics is increasing, medical students receive comparatively little training in the skills needed to assess environmental health hazards (Chivian et al. 1993, 5). In the training of biologists, molecular and genetic issues get an ever larger share while systematic knowledge about the living world has a decreasing place. The linguistic simplifications concerning genes encourage such tendencies; studies into the "genetic basis" of health or behavior, for example, suggest that the basic or most fundamental thing concerning health or behavior is genetic. The same is true for the expression "gene for x." Further, there is the influence of the Weissmannian doctrine of the universal separation of the germ line, refuted though it is. According to this simplistic doctrine, causal arrows can be drawn from the germ cells to the body, but not the other way around. The current weight placed on genes in understanding biological causation partly builds on this tradition of simplistically drawing causal arrows (see Griesemer 1994).

The allocation of scientific attention influences the solutions that are available to existing problems, as well as the identification and characterization of new problems. Genetic aspects of diseases are increasingly identified as important problems that need attention. Applications of genetic developmental biology are predominantly foreseen in this sphere of medical genetics. It has been noticed that the factors that give embryological tissues their remarkable capacity of growth and differentiation might provide useful medical treatments: "If similar capacities could be bestowed at will on the adult human, many diseases could be cured simply by growing new tissues to replace damaged ones" (Nowak 1994, 567). Embryonic growth and differentiation factors such as bone-growth promoting factor, that might be helpful for healing bone fractures, are now rapidly becoming medically and commercially interesting.

The Council for Responsible Genetics (1990) has warned against the human genome project, criticizing the reductionist view of causality that detracts from other biological processes and from social factors: "Focusing on genes as the cause of our various problems will make it more difficult to enact appropriate social policies." The Council comments on the comparatively large amounts of money spent on mapping and studying genes by saying that "the genome project vastly exaggerates the importance of genes—especially at this time, when a deteriorating environment and economy make it increasingly difficult for most people to live healthful lives" (p. 4). This theme of the increasing geneticizing of health and disease is prominent in the volume *Are Genes Us* (1994), edited by Carl Cranor. For example, Evelyn Fox Keller argues in her paper in this book that though some of the experimental studies in the context of the human genome project are inevitably bound to undermine the simplistic notions on which they rely, the project certainly encourages the trend that genes are now often seen not only as causing disease, but as defining disease.

Ruth Hubbard has developed this line of thought in more detail. In *Exploding the Gene Myth* (1993), a book she wrote with Elijah Wald, she intends to place genes in their proper context, identifying old and new forms of eugenics and medical onesidedness as results of an overemphasis on genes. Biomedical scientific attention is overly reductionistic and excessively aimed at individual causes of health, which draws attention to affected individuals and away from systemic conditions, she writes. There should be a balance between public health measures and individual health care; the absence of such a balance shows up in onesided policies such as the screening of workers for their individual suspectibilities instead of the cleaning up of factories (p. 61). The human genome project worsens the already existing imbalance in scientific and social attention, according to Hubbard, and she is profoundly skeptical about the ethical program in the margin of this project (ELSI), because this will probably not affect the direction of scientific research itself.

Though Hubbard assumes, rightly I think, that ethics usually does not touch upon scientific research, her concerns are in fact profoundly moral. She worries about the social consequences of scientific onesidedness; in other words, her approach fits in with what I have called "ethics of attention." Her treatment suggests balanced attention as a normative ideal for science, which may indeed be the natural ideal for a normative treatment of scientific attention. Though it is bound to be controversial in many cases what constitutes balance, for example because it will be theory-dependent, this is no reason not to try and develop this notion in more depth. There may appear to be many cases in which it is not so very hard to reach agreement after all.

How can a balance of attention, whatever its content, be reached, starting from an unbalanced situation? Critically exposing the narrow focus on overdeveloped areas is a natural strategy, but it also has its dangers. It keeps those areas at the center of attention. What about neglected issues? Positivity bias ensures that neglect is an easily neglected harm, since nothing positively bad is done. To counterbalance this, attention should also be paid to what is missing; in other words, neglected areas should be located and developed. As long as attention remains focused on the role of genes in development, the neglect of the environmental embeddedness of development will continue in various forms.

The "human genome diversity project" has been set up to analyze and compare DNA samples of hundreds of different ethnic groups. Among these are aboriginal groups that are on the verge of extinction. The project has met with criticism because it "swoops in," collects the blood of these peoples, and then leaves them to their fate (Lewin 1993, 25). Science thus leaves the impression that it is only interested in its own favorite goals and kinds of knowledge, not in the well-being of its subjects of research.

Many animals are in the same position as the aboriginal groups just mentioned. In a world in which natural environments are deteriorating, what contributions can we expect from a developmental biology that focuses on genes? One possible answer is that we might end up knowing everything about the genes of organisms that no longer exist because of environmental problems.

Different approaches to development imply different starting points to answer questions about what it means and takes to live and flourish, and what can possibly go wrong. Overall approaches to development matter because they determine what kind of data are important. The data themselves are also important. When you say that animals are essentially developing in an environment, but no data exist on how specifically environmental causes are involved in development, it is hard to give this view a place in discussions. In-principle acknowledgment of the environment may then be combined with continued gene-centric research.

The lack of understanding of environmental influences in development has made it possible that dramatic damage has gone undetected for a long period: the right questions were not asked. Toxicological research makes it now increasingly clear that hormonelike effects of pesticides cause severe developmental problems in various animals, such as gulls and terns, alligators, frogs, fishes and birds (Fry and Toone 1981; Colborn and Clement 1992; Stone 1992,

1994; Mestel 1994; Blaustein et al 1994; Culotta 1995). The view that development is controlled by a developmental program does not direct attention to such phenomena. In my view, development had better be defined, approached, and studied in a broader way, that is to say, in an ecological way. The knowledge thus generated may be helpful in finding solutions for the severe problems in the development of many animals that have their origins in the environment.

Figure 6.1
Arctic tern, *Sterna paradisaea*

Notes

Chapter 1

1. Apart from pragmatism, he sees "formism," "mechanism," and "organicism" or "absolute idealism" as fundamental perspectives. Formism or platonic idealism is a categorizing world view that is based on the root metaphor of similarity and that emphasizes classification; it has yielded the theory of types. Mechanism has the machine as its root metaphor. Finally (and weakest of the four, says Pepper) there is organicism or absolute idealism. Its root metaphor is the integration of historic events.

Chapter 2

1. A *Drosophila* egg is connected with the cytoplasm of a number of nurse cells at the future anterior (head) end. From these cells, proteins, ribosomes, and mRNA are transported into the egg. Development starts with a number of cleavages of the nucleus. This yields a so called syncytial blastoderm, which is to say that there are no cell walls apart from the outer cell wall: all the nuclei are contained in the same cytoplasm. The nuclei all move to the periphery, just under the cell wall, encircling the central yolk of the egg. During the ninth division, about five nuclei reach the future posterior pole, and become enclosed in a cell wall. These cells later give rise to the gametes in the adult. After thirteen cycles of division, when there are about 6,000 nuclei, cell walls are formed. At the fourteenth round of division, transcription, which has been low up to then, increases enormously. At the same time, the cells become motile and gastrulation begins (see Gilbert 1994, 531).

2. The distinction resembles Schmalhausen's distinction between autoregulatory and dependent morphogenesis (Schmalhausen 1949, 34–35).

3. See Van der Weele (1993b) for an earlier treatment of this distinction. Frankel (1990), Smith (1993) and Van der Weele (1993a) distinguish two perspectives: a genetic or neo-Darwinian versus a structuralist one.

4. Other examples of work in this tradition are Meinhardt (1982), Oster and Alberch (1982), Oster et al. (1988), Oster and Murray (1989), Murray (1989), Newman and Comper (1990), Kulesa et al. (1994).

5. Slack (1991) discusses mathematical models, embryological experiments as well as developmental genetics with the aim to bring them together.

6. This field approach should not be confused with Sheldrake's, who writes about fields of morphic resonance. The primary difference, says Goodwin, is that Sheldrake's fields are nonphysical, while he himself is talking about fields defined by physical and chemical processes (Goodwin 1994a, 88).

7. For (mathematical) treatments of *Acetabularia* growth see also Goodwin and Trainor (1985) and Brière and Goodwin (1988).

8. Let me briefly mention Goodwin's relation with biological and philosophical traditions. The structuralist approach, according to Goodwin, aims at a "rational morphology" and a "science of qualities"; terms that are meant to establish a link with Goethe and other morphologists of the late eighteenth and early nineteenth century (for example, see Webster and Goodwin 1981, 1982; Goodwin 1989a, 1990b, 1994a, b). Organisms should be seen as integrated wholes that express their essential nature. E. M. Barth (1977) has analyzed what may be called "the logic of quantity and quality" (which associates rationality with quality, and values them much higher than quantity) in German philosophy of Goethe's time. This old philosophical paradigm is very problematic from a logical point of view.

Fascinating questions can be asked concerning the origin, meaning, and consequences of the parallels between Goodwin's view and this old logic, but I will not go into them. According to Goodwin (personal communication), the reference to Goethean holism is intentional, but a revival of the logic of quantity and quality is not.

9. Several other authors fit in with this approach; see Fogel (1993), Gray (1992), Griffiths and Gray (1994), Lickliter (1990), Lickliter and Banker (1994), Gottlieb (1992), Johnston and Gottlieb (1990), and further back, for example Levins and Lewontin (1985), Rose et al. (1984), Bateson (1983) and Lehrman (1970).

10. Sydney Brenner, interviewed by Roger Lewin, makes a similar suggestion: "I'm not sure that there necessarily is anything more to understand than what it is," he says, adding that perhaps simple descriptions of how organisms are put together may yield the most profound insights (Lewin 1984). For explicit criticism on the notion of programming, see also S. A. Newman (1988, 1989); Stent (1985); Sober (1985). Sober writes: "A fly that is smashed by a fly swatter is not executing a conditional program that says "Die if hit by a large object." Nor is it following any sort of program either. Likewise, I see no reason to think that flies at the beginning of the experiment (concerning genetic assimilation, CvdW) encoded the instruction "develop bithorax if and only if an ether stimulus is present" (Sober 1985, 205).

11. As Susan Oyama has remarked (personal communication), part of the confusion may have to do with associations between issues of causation and the active-passive dichotomy. This dichotomy is indeed very influential in thought and language about causes and would deserve a thorough analysis. I will not undertake such analysis here, but let me quote John Stuart Mill, who wrote: "In most cases of causation a distinction is commonly drawn between something which acts, and some other thing which is acted upon; between an *agent* and a *patient*. Both of these, it would be universally allowed, are conditions of the phenomenon; but it would be thought absurd to call the latter the cause, that title being reserved for the former. The distinction, however, vanishes on examination, or rather is found to be only verbal; arising from an incident of mere expression, namely, that the object said to be acted upon, and which is considered as the scene in which the effect takes place, is commonly included in the phrase by which the effect is spoken of, so that if it were also reckoned as part of the cause, the seeming incongruity would arise of its being supposed to cause itself (Mill 1973, 334-335 (book III)).

12. This view has a parallel in "constructivist" approaches in social studies of science, where a reluctance to allow a somehow privileged status for any partial perspective also exists. Woolgar (1988), for example, seeks to subvert all representation, scientific and otherwise, by methodic reflexivity.

Chapter 3

1. Nor is Hempel committed to the view that there can only be one complete explanation. At least, this is the view of Ruben (1993, 3); but there is room for different interpretations in this respect.

2. Van Fraassen, too, refers to "the causal net" that events are enmeshed in. Explanations cannot refer to the causal net as a whole, or even to the part that has to do with a specific question, because it is too large: "To describe the whole causal net in any connected region, however small, is in almost every case impossible" (Van Fraassen 1980, 124). Although he defines the causal net as "whatever structure of relations science describes," he also writes that "explanatory factors are to be chosen from a range of factors which are (or which the scientific theory lists as) objectively relevant in certain special ways" (1980, 126).

3. "Subjectivity" is a haunting ghost for defenders of ontic approaches to causation. Paul Humphreys (1989) is another of those defenders. On the basis of his view that pragmatics implies subjectivism, he is arguing against it in his book *The Chances of Explanation*, where he develops an ontological, antipragmatic view of causation. He defines pragmatics in a subjectivistic and linguistic way. Pragmatics is about the meaning of terms, and it refers to individuals, their purposes and motives. It seems that for Humphreys there are only two possibilities: causal explanations either express objective mechanisms or subjective preferences. A further difference with my treatment is that he defines pragmatics in a linguistic way, while I also include theoretical choices.

4. For overviews see Hesslow (1988) and Gifford (1990).

5. Mackie remarks that the complete set of relevant factors, together with a function which shows in which way they are relevant, may be called the complete cause (Mackie 1974, 149).

6. Lewontin's work in population genetics, though, is mathematical.

7. An example of the distinction between general "'views" and "restricted questions" can be found in Lewontin's criticism of claims surrounding the human genome project: "The problem with this story is that although it is correct in its detailed molecular description, it is wrong in what it claims to explain. First, DNA is not self-reproducing, second, it makes nothing, and third, organisms are not determined by it" (Lewontin 1992a).

8. Toward the end of his book *Adaptation and Environment*, Brandon (1990) deals with the problem of ideally complete explanations and gives an example of an evolutionary explanation that he considers to be reasonably complete. Completeness for Brandon is thus relative to a theory: the theory of evolution. Is the theory he discusses complete about the causes of the phenomenon to be explained? Certainly not; it is a specific theory

about historical causes that disregards, among other things, developmental and physiological causes. Besides, controversial issues in evolutionary theory, for example concerning levels of selection, render it controversial what should and shouldn't be included in a complete explanation. Clearly, a "complete" explanation in Brandon's sense comes nowhere near the Salmon/ Railton ideally complete explanatory text or theory. In other words, not even the most comprehensive theory in biology represents a complete approach.

9. Nicolas Rescher (1984) also deals with the theoretical context as if it were a single unified whole. For Rescher, as for Van Fraassen, a "complete" explanation is a satisfactory or adequate answer to a question. Now scientific questions are not isolated arbitrary whims; they contain presuppositions: "A presupposition of a question is a thesis (or proposition) that is inherent in each of its possible fully explicit answers. For example, " 'What is the melting point of lead?' has innumerably many fully explicit answers, all taking the general form 'm °C is the melting point of lead.' All of these imply that lead indeed has a (fixed and stable) melting point" (Rescher 1984, 19). That a question cannot be separated from its presuppositions is one of the reasons why knowledge cannot be separated from its historical context. When knowledge changes, the body of fruitful questions changes. If lead were not known to have a melting point, it would be pointless to ask for it. In Rescher's view, the body of knowledge associated with a specific time, generates the "question agenda" of that time: the set of questions that can appropriately be posed with respect to that knowledge. Rescher clearly sees the "body of knowledge" of a certain time as a unified whole.

10. The case concerns the evolutionary explanation of female orgasm. S. J. Gould has challenged the appropriateness of an adaptationist explanation, arguing that female orgasm is more likely to be a developmental by-product of the homologous male trait. Alcock, on the other hand, has suggested a possible adaptive function for clitoral orgasm. Gould and Alcock are both talking about the causal history responsible for clitoral orgasm; "they do not disagree upon the question, but rather upon what will constitute evidence for the various competing causal answers" (Mitchell 1992, 138).

Chapter 4

1. See on this period also Russell (1982 [1916]).

2. It has also met with other kinds of criticism. Richard Francis for example has denied that ultimate causes are causes at all. Ultimate explanations, according to him, are concerned with certain types of effects, namely, adaptations. Their existence is inferred from functional analysis; hence, ultimate explanations are functional explanations, which do not refer to causes, but to the way things hang together (Francis 1990).

3. J. Antonovics and P. H. van Tienderen (1991) have complained that theoretical contexts are often left vague for lack of a null model. Confusing debates about constraints are the result. Explicit stating of the null model obviates the need to use the term constraint, they say. In their view, "a temporary moratorium on constraint usage may contribute to clearer thinking within the field." See also Van Tienderen and Antonovics (1994).

4. Waddington, who introduced the issue of genetic assimilation, was not a neo-Lamarckian either. For Waddington, genetic assimilation involved selection on previously hidden genetic variation (previously unexpressed genes being expressed in new environments); see Waddington (1953), and also Hall (1992) and Rollo (1994).

5. Subsequently, Hall discusses the concept of genetic assimilation and canalization in the work of Waddington and others, as well as attempts by Atchley and himself to initiate a quantitative genetics model for morphological change in development and evolution. One of the difficulties for such a model is that the basic units are hard to define: how to quantify shape? (Hall 1992, 176; Atchley 1987; Atchley and Hall 1991).

6. The other three are: (a) the received view, which sees evolution as resulting from competition between organisms, (b) the gene's eye view, in which evolution results from competition between genes that affect interactors, (c) the extended replicator view, in which genes are not the only replicators (this is their own view). They agree with developmental systems theorists on three points: (1) the genome of a developing organism is not causally sufficient for its development; (2) genes are not the only bridge between the generations and (3) the skin is not an important causal boundary.

Chapter 5

1. See, for example, Rendel (1962), Scharloo (1987, 1988). The morphogenetic substances have normal distributions in these models, due to genetic, environmental and stochastic variation.

2. For critical comments on this term, see Oyama (1981) and Rose (1995).

Chapter 6

1. The effects of genetic differences depend on the environment. For example, for the development of PKU (phenylketonuria), not only certain genes are needed, but also a womb of the right kind for the fetus to develop its traits, including its inability to metabolize phenylalanine. This trait in turn does not in itself lead to PKU, but only in combination with a normal diet.

2. Dawkins writes: "If society systematically trains children without penises to knit and play with dolls, and trains children with penises to play with guns and toy soldiers, any resulting differences in male and female preferences are strictly speaking genetically determined differences! They are determined, through the medium of societal custom, by the fact of possession or nonpossession of a penis, and that is determined (. . .) by sex chromosomes. (. . .) All genetic causes have to work in the context of an environment of some kind. If a genetic sex difference makes itself felt through the medium of a sex-biased education system, it is still a genetic difference" (Dawkins 1982, 12).

3. In response to Sober (1992). See also Mitchell (1987).

4. Though there is a potentially troublesome problem of infinite regress: for every proximate cause, an even more proximate cause may exist. See Sober (1992), who argues that it is not plausible that going closer and closer to the effect always yields greater explanatory power.

5. I owe this example to Wim van der Steen.

6. In one place Lewontin (1992b) goes so far as to say that causes can only be social, while illness-producing factors like asbestos are only the "agents" of social causes. The underlying motivation may be to do justice to the social embeddedness of illness, and to implement the view that contexts should not be defined too narrowly. But this particular distinction between causes and agents does not seem to be fruitful, and even Lewontin himself is not consistent in the use of it, as he freely talks about all kinds of causes in other places. I think there are good reasons to stick to the more pragmatic and open approach to causation.

7. This is a different kind of integration between evolution and development than was discussed in Chapter 4. It does not emphasize the relevance of development for evolution, but the relevance of evolution for development. Atkinson (1992b) mentions both paths to integration.

References

Alberts, B., D. Bray, J. Lewis, M. Raff, K. Roberts, and J. D. Watson. (1989). *Molecular Biology of the Cell*. (2d ed.). New York: Garland.

Allen, G. E. (1978). *Life Science in the Twentieth Century*. Cambridge: Cambridge U.P.

Allen, G. E. (1985). T. H. Morgan and the split between embryology and genetics, 1910–35. In T. J. Horder, J. A. Witkowski, and C. C. Wylie (eds.), *A History of Embryology* 113–46. Cambridge: Cambridge U.P.

Amundson, R. (1994). Two concepts of constraint: Adaptationism and the challenge from developmental biology. *Philosophy of Science* 61:556–78.

Antonovics, J., and P. H. van Tienderen. (1991). Ontoecogenophyloconstraints? The chaos of constraint terminology. *TREE* 6:166–68.

Atchley, W. R. (1987). Developmental quantitative genetics and the evolution of ontogenies. *Evolution* 41:316–30.

Atchley, W. R., and B. K. Hall. (1991). A model for development and evolution of complex morphological structures. *Biological Reviews* 66:101–57.

Atkinson, B. G., T. Cunningham, R. L. Dean, and M. Somerville. (1983). Comparison of the effects of heat shock and metal-ion stress on gene expression in cells undergoing myogenesis. *Canadian Journal of Biochemistry and Cell Biology* 61:404–13.

Atkinson, J. W. (1992a). Conceptual issues in the reunion of development and evolution. *Synthese* 91:93–110.

Atkinson, J. W. (1992b). Development and macroevolution: Introduction to the symposium. *American Zoologist* 32:103–5.

Barth, E. M. (1974). *The Logic of the Articles in Traditional Philosophy.* Dordrecht: Reidel.

Barth, E. M. (1977). The logical paradigm in dialectical philosophy and science. *Erkenntnis* 11:291–322.

Barth, E. M., and E. C. W. Krabbe. (1982). *From Axiom to Dialogue: A Philosophical Study of Logics and Argumentation.* Berlin: Walter de Gruyter.

Bateson, P. (1983). Genes, environment and the development of behaviour. In T. R. Halliday and P. J. B. Slater (eds.), 52–81. *Genes, Development and Learning.* Oxford: Blackwell.

Beardsley, T. (1991). Smart genes. *Scientific American* (August): 73–81.

Beatty, J. (1994). The proximate/ultimate distinction in the multiple careers of Ernst Mayr. *Biology and Philosophy* 9:333–56.

Beck, S. D. (1980). *Insect Photoperiodism.* New York: Academic Press.

Bellah, R. N., R. Madsen, W. M. Sullivan, A. Swidler, and S. M. Tipton. (1985). *Habits of the Heart: Individualism and Commitment in American Life.* Berkeley: University of California Press.

Bestor, T. H., V. I. Chandler, and A. P. Feinberg. (1994). Epigenetic effects in eukaryotic gene expression. *Developmental Genetics* 15:458–62.

Bink, F. A. (1992). *Ecologische Atlas van de Dagvlinders van Noordwest-Europa.* Haarlem: Schuyt.

Blaustein, A. R., D. B. Wake, and W. P. Sousa. (1994). Amphibian declines: Judging stability, persistence, and suspectibility of populations to local and global extinctions. *Conservation Biology* 8 (1):60–71.

Bohm, D., and F. D. Peat. (1987). *Science, Order, and Creativity.* London: Routledge.

Bond, U., and M. J. Schlesinger. (1987). Heat-shock proteins and development. *Advances in Genetics* 24:1–29.

Bournias-Vardiabasis, N., and C. H. Buzin. (1987). Altered differentiation and induction of heat shock proteins in *Drosophila* embryonic cells associated with teratogen treatment. In J. A. McLachlan, R. M. Pratt and C. L. Markert (eds.), *Developmental Toxicology: Mechanisms and Risk* 3–16. Cold Spring Harbor: CSH-Laboratory.

Bowers, W. S., T. Ohta, J. S. Cleere, and P. A. Marsella. (1976). Discovery of insect anti-juvenile hormones in plants. *Science* 193:542–47.

Brakefield, P. M., and N. Reitsma. (1991). Phenotypic plasticity, seasonal climate and the population biology of *Bicyclus* butterflies (Satyridae) in Malawi. *Ecological Entomology* 16:291–303.

Brandon, R. N. (1990). *Adaptation and Environment*. Princeton: Princeton U.P.

Brandon, R. N., J. Antonovics, R. Burian, S. Carson, G. Cooper, P. S. Davies, C. Horvath, B. D. Mishler, R. C. Richardson, K. Smith, and P. Thrall. (1994). Discussion: Sober on Brandon on screening-off and the levels of selection. *Philosophy of Science* 61:475–86.

Braun, A. G. (1987). Teratogen metabolism. In J. A. McLachlan, R. M. Pratt, and C. L. Markert (eds.), *Developmental Toxicology: Mechanisms and Risk* 17–28. Cold Spring Harbor: CSH-laboratory.

Brian, M. V. (1980). Social control over sex and caste in bees, wasps and ants. *Biological Reviews* 55:379–415.

Brière, C., and B. Goodwin. (1988). Geometry and dynamics of tip morphogenesis in *Acetabularia*. *Journal of Theoretical Biology* 131:461–75.

Bronfenbrenner, U. (1979). *The Ecology of Human Development*. Cambridge (Mass.): Harvard U.P.

Bull, J. J. (1980). Sex determination in reptiles. *The Quarterly Review of Biology* 55:3–21.

Bull, J. J. (1983). *Evolution of Sex Determining Mechanisms*. Menlo Park: Benjamin/Cummings.

Bull, J. J., and R. C. Vogt. (1979). Temperature-dependent sex determination in turtles. *Science* 206:1186–88.

Buss, L. W. (1987). *The Evolution of Individuality*. Princeton: Princeton U.P.

Cartwright, N. (1983). *How the Laws of Physics Lie*. Oxford: Clarendon Press.

Chivian, M. D., M. McCally, H. Hu, and A. Haines. (1993). *Critical Condition; Human Health and the Environment*. Cambridge (Mass.): MIT Press.

Colborn, T., and C. Clement. (1992). *Chemically-Induced Alterations in Sexual and Functional Development: The Wildlife/Human Connection*. Princeton: Princeton Scientific Publishing Co.

Council for Responsible Genetics. (1990). *Position Paper on Genetic Discrimination*. Boston: Council for Responsible Genetics.

Cranor, C. F. (1994). *Are Genes Us? The Social Consequences of the New Genetics.* New Brunswick: Rutgers U.P.

Crowl, T. A., and A. P. Covich. (1990). Predator-induced life-history shifts in a freshwater snail. *Science* 247:949–50.

Culotta, E. (1995). St. Louis meeting showcases "creature features." *Science* 267:330–31.

Darwin, C. (1964 [1859]). *On the Origin of Species.* Cambridge (Mass.): Harvard U.P.

Dawkins, R. (1982). *The Extended Phenotype.* Oxford: Oxford U.P.

Day, S. (1990). Genes that control genes. *New Scientist* (3 November), Inside Science: 1–4.

De Beer, G. (1958 [1940]). *Embryos and Ancestors.* Oxford: Clarendon Press.

De Pomerai, D. (1990). *From Gene to Animal* (2nd ed.). Cambridge: Cambridge U.P.

Deeming, D. C., and M. W. J. Ferguson. (1988). Environmental regulation of sex determination in reptiles. *Philosophical Transactions of the Royal Society London* B 322:19–39.

Depew, D. J., and B. H. Weber. (1995). *Darwinism Evolving.* Cambridge (Mass.): MIT Press.

Dodson, S. (1989). Predator-induced reaction norms. *BioScience* 39:447–52.

Doucet, P., and P. B. Sloep. (1992). *Mathematical Modelling in the Life Sciences.* New York: Ellis Horwood.

Douglas, K. (1993). The great folic acid scandal. *New Scientist* (17 July):24–25.

Evans, J. S. B. T. (1989). *Bias in Human Reasoning: Causes and Consequences.* Hove: Lawrence Erlbaum Associates.

Fetzer, J. H. (1987). Wesley Salmon's "Scientific explanation and the causal structure of the world." *Philosophy of Science* 54:597–610.

Fogel, A. (1993). *Developing through Relationships.* New York: Harvester Wheatsheaf.

Forst, S., and M. Inouye. (1988). Environmentally regulated gene expression for membrane proteins in *Escherichia coli. Annual Review of Cell Biology* 4:21–42.

Fox, G. A. (1992). Epidemiological and pathobiological evidence of contaminant-induced alterations in sexual development in free-living wildlife. In T. Colborn and C. Clement (eds.), *Chemically-Induced Alterations in Sexual and Functional Development: The Wildlife/Human Connection*. Princeton: Princeton Scientific Publishing Co.

Francis, R. C. (1990). Causes, proximate and ultimate. *Biology and Philosophy* 5:401–15.

Frankel, J. (1990). The evolution of development (book review). *Evolution* 44:465–67.

Fry, M. D., and C. K. Toone. (1981). DDT-induced feminization of gull embryos. *Science* 213:922–24.

Garfinkel, A. (1981). *Forms of Explanation*. New Haven: Yale U.P.

Gehring, W. J. (1987). Homeo boxes in the study of development. *Science* 236:1245–52.

Gifford, F. (1990). Genetic traits. *Biology and Philosophy* 5:327–47.

Gilbert, J. J. (1966). Rotifer ecology and embryological induction. *Science* 151:1234–37.

Gilbert, J. J. (1980). Female polymorphism and sexual reproduction in the rotifer *Asplanchna*: Evolution of their relationship and control by dietary tocopherol. *The American Naturalist* 116:409–31.

Gilbert, S. F. (1991a). Epigenetic landscaping: Waddington's use of cell fate bifurcation diagrams. *Biology and Philosophy* 6:135–54.

Gilbert, S. F. (1991b). Induction and the origins of developmental genetics. In S. F. Gilbert (ed.), *A Conceptual History of Modern Embryology*. 181–206. New York: Plenum Press.

Gilbert, S. F. (1991c). The role of embryonic induction in creating self. In A. I. Tauber (ed.), *Organism and the Origins of Self* 341–60. Dordrecht: Kluwer.

Gilbert, S. F. (1992). Cells in search of community: Critiques on Weismannism and selectable units in ontogeny. *Biology and Philosophy* 7:473–87.

Gilbert, S. F. (1994). *Developmental Biology* (4th ed.). Sunderland: Sinauer.

Goodwin, B. C. (1984a). Changing from an evolutionary to a generative paradigm in biology. In J. W. Pollard (ed.), *Evolutionary Theory: Paths into the Future* 99–120. Chichester: John Wiley and Sons Ltd.

Goodwin, B. C. (1984b). A relational or field theory of reproduction and its evolutionary implications. In M. W. Ho and P. T. Saunders (eds.), *Beyond Neo-Darwinism* 219–41. New York: Academic Press.

Goodwin, B. C. (1985). What are the causes of morphogenesis? *BioEssays* 3:32–36.

Goodwin, B. C. (1987). A science of qualities. In B. J. Hiley and F. D. Peat (eds.): *Quantum Implications. Essays in Honour of David Bohm* 328–37. New York: Routledge and Kegan Paul.

Goodwin, B. C. (1988a). Morphogenesis and heredity. In M. W. Ho and S. W. Fox (eds.), *Evolutionary Processes and Metaphors* 145–62. Chichester: John Wiley and Sons Ltd.

Goodwin, B. C. (1988b). Problems and prospects in morphogenesis. *Experientia* 44:633–37.

Goodwin, B. C. (1989a). Evolution and the generative order. In B. C. Goodwin and P. Saunders (eds.), *Theoretical Biology.* 89–100. Edinburgh: Edinburgh U.P.

Goodwin, B. C. (1989b). A structuralist research programme in developmental biology. In B. C. Goodwin, A. Sibatani and G. Webster (eds.), *Dynamic Structures in Biology* 49–61. Edinburgh: Edinburgh U.P.

Goodwin, B. C. (1990a). The causes of biological form. In G. Butterworth and P. Bryant (eds.), *Causes of Development* 49–63. New York: Harvester Wheatsheaf.

Goodwin, B. C. (1990b). Structuralism in biology. *Scientific Progress* 74:227–44.

Goodwin, B. C. (1994a). *How the Leopard Changed its Spots.* London: Weidenfeld and Nicolson.

Goodwin, B. C. (1994b). Towards a science of qualities. In W. Harman and J. Clark (eds.), *New Metaphysical Foundations of Modern Science* 215–50. Sausalito (Calif.): Institute of Noetic Sciences.

Goodwin, B. C., S. Kauffman, and J. D. Murray. (1993). Is morphogenesis an intrinsically robust process? *Journal of Theoretical Biology* 163:135–44.

Goodwin, B. C., and L. E. H. Trainor. (1985). Tip and whorl morphogenesis in *Acetabularia* by calcium-regulated strains. *Journal of Theoretical Biology* 117:79–106.

Gottlieb, G. (1992). *Individual Development and Evolution*. New York & Oxford: Oxford U.P.

Gould, S. J. (1977). *Ontogeny and Phylogeny*. Cambridge (Mass.): Harvard U.P.

Gould, S. J. (1985). *The Flamingo's Smile*. London: Penguin.

Gould, S. J. (1989). A developmental constraint in *Cerion*, with comments on the definition and interpretation of constraint in evolution. *Evolution* 43:516–39.

Gould, S. J. (1991 [1989]). *Wonderful Life*. London: Penguin.

Gould, S. J. (1992). Ontogeny and phylogeny—revisited and united. *BioEssays* 14:275–79.

Gould, S. J., and R. C. Lewontin. (1979). The spandrels of San Marco and the Panglossian paradigm: A critique of the adaptationist programme. *Proceedings of the Royal Society London* B 205:581–98.

Gray, R. (1992). Death of the gene: developmental systems strike back. In P. Griffiths (ed.), *Trees of Life: Essays on the Philosophy of Biology* 165–209. Dordrecht: Kluwer.

Greene, E. (1989). A diet-induced developmental polymorphism in a caterpillar. *Science* 243:643–46.

Griesemer, J. R. (1994). Tools for talking: Human nature, Weismannism, and the interpretation of genetic information. In C. F. Cranor (ed.) *Are Genes Us?* 69–88. New Brunswick: Rutgers U.P.

Griffiths, A. J. F., J. H. Miller, D. T. Suzuki, R. C. Lewontin, and W. C. Gelbart. (1993). *An Introduction to Genetic Analysis*. New York: Freeman.

Griffiths, P. E., and R. D. Gray. (1994). Developmental systems and evolutionary explanation. *Journal of Philosophy* 91:277–304.

Gupta, A. P., and R. C. Lewontin. (1982). A study of reaction norms in natural populations of *Drosophila pseudoobscura*. *Evolution* 36:934–48.

Hall, B. K. (1983). Epigenetic control in development and evolution. In B. C. Goodwin, N. Holder, and C. C. Wylie (eds.), *Development and Evolution* 353–80. Cambridge: Cambridge U.P.

Hall, B. K. (1992). *Evolutionary Developmental Biology*. London: Chapman & Hall.

Hamburger, V. (1980). Embryology and the modern synthesis in evolutionary theory. In E. Mayr and W. B. Provine (eds.), *The Evolutionary Synthesis* 97–112. Cambridge (Mass.): Harvard U.P.

Hanson, N. R. (1958). *Patterns of Discovery*. Cambridge: Cambridge U.P.

Haraway, D. J. (1976). *Crystals, Fabrics and Fields*. New Haven: Yale U.P.

Hart, H. L. A., and T. Honoré. (1985 [1959]). *Causation in the Law*. Oxford: Clarendon Press.

Havel, J. E., and S. I. Dodson. (1984). *Chaoborus* predation on typical and spined morphs of *Daphnia pulex*: Behavioral observations. *Limnology and Oceanography* 29:487–94.

Heikkila, J. J. (1993a). Heat shock gene expression and development. I. An overview of fungal, plant, and poikilothermic animal developmental systems. *Developmental Genetics* 14:1–5.

Heikkila, J. J. (1993b). Heat shock gene expression and development. II. An overview of mammalian and avian developmental systems. *Developmental Genetics* 14:87–91.

Held, L. I. (1992). *Models for Embryonic Periodicity*. Basel: Karger.

Hempel, C. G. (1962). Explanation in science and history. In R. G. Colodny (ed.), *Frontiers of Science and Philosophy*. London: Allen & Unwin (Reprinted in Ruben 1993).

Hempel, C. G. (1965). *Aspects of Scientific Explanation*. New York: Free Press.

Hempel, C. G., and P. Oppenheim. (1936). *Der Typusbegriff im Lichte der Neuen Logik*. Leiden: Sijthoff.

Hempel, C. G., and P. Oppenheim. (1948). Studies in the logic of explanation. *Philosophy of Science* 15:135–75.

Hertwig, O. (1894). *Zeit- und Streitfragen der Biologie Heft I: Präformation oder Epigenese?* Jena: Gustav Fischer Verlag.

Hesslow, G. (1984). What is a genetic disease? On the relative importance of causes. In L. Nordenfelt and B. I. B. Lindahl (eds.), *Health, Disease, and Causal Explanations in Medicine*. Dordrecht: Reidel.

Hesslow, G. (1988). The problem of causal selection. In D. J. Hilton (ed.), *Contemporary Science and Natural Explanation* 11–32. New York: New York U.P.

Highsmith, R. C. (1982). Induced settlement and metamorphosis of sand dollar (*Dendraster excentricus*) larvae in predator-free sites: Adult sand dollar beds. *Ecology* 63 (2):329–37.

Hodges, W. (1978). *Logic*. Harmondsworth: Penguin.

Hofstadter, D. R. (1985). The genetic code: arbitrary? In D. R. Hofstadter, *Metamagical Themas* 671–99. London: Penguin.

Holliday, R. (1987). The inheritance of epigenetic defects. *Science* 238:163–70.

Holliday, R. (1990). Mechanisms for the control of gene activity during development. *Biological Reviews* 65:431–71.

Holliday, R. (1994). Epigenetics: An overview. *Developmental Genetics* 15:453–57.

Horder, T. J. (1994). Partial truths: A review of the use of concepts in the evolutionary sciences. In R. W. Scotland, D. J. Siebert and D. M. Williams (eds.), *Models in Phylogeny Reconstruction*. Special issue of The Systematics Association. Oxford: Clarendon Press.

Hubbard, R., and E. Wald. (1993). *Exploding the Gene Myth*. Boston: Beacon Press.

Humphreys, P. (1989). *The Chances of Explanation*. Princeton: Princeton U.P.

Infante, A. A., D. Infante, and J. Rimland. (1993). Ferritin gene expression is developmentally regulated and induced by heat shock in sea urchin embryos. *Developmental Genetics* 14:58–68.

Jablonka, E., and M. J. Lamb. (1989). The inheritance of acquired epigenetic variations. *Journal of Theoretical Biology* 139:69–83.

Jablonka, E., M. Lachmann, and M. J. Lamb. (1992). Evidence, mechanisms and models for the inheritance of acquired characters. *Journal of Theoretical Biology* 158:245–68.

Jacob, F., and J. Monod. (1961). Genetic regulatory mechanisms in the synthesis of proteins. *Journal of Molecular Biology* 3:318–56.

Janzen, F. J., and G. L. Paukstis. (1991). Environmental sex determination in reptiles: Ecology, evolution, and experimental design. *The Quarterly Review of Biology* 66:149–79.

Jeffery, W. R., and B. J. Swalla. (1992). Evolution of alternate modes of development in Ascidians. *BioEssays* 14:219–26.

Johnston, T. D., and G. Gottlieb. (1990). Neophenogenesis: A developmental theory of phenotypic evolution. *Journal of Theoretical Biology* 147:471–95.

Kacser, H. (1960). Kinetic models of development and heredity. In J. W. L. Beament (ed.), *Models and Analogues in Biology*. Cambridge: Cambridge U.P.

Kahn, P. (1994). Zebrafish hit the big time. *Science* 264:904–5.

Kauffman, S. A. (1983). Developmental constraints: Internal factors in evolution. In B. C. Goodwin, N. Holder and C. C. Wylie (eds.), *Development and Evolution* 195–226. Cambridge: Cambridge U.P.

Kauffman, S. A. (1993). *The Origins of Order*. New York & Oxford: Oxford U.P.

Keeton, W. T., and J. L. Gould. (1986). *Biological Science* (4th ed.). New York: Norton.

Keller, E. Fox (1992). *Secrets of Life, Secrets of Death*. New York: Routledge.

Keller, E. Fox (1994). Master molecules. In C. F. Cranor (ed.) *Are Genes Us?* 89–98. New Brunswick: Rutgers U.P.

Kim, K. S., Y. K. Kim, F. Naftolin, and C. L. Markert. (1987). Synthesis of stress proteins during normal and stressed development of mouse embryos. In J. A. McLachlan, R. M. Pratt and C. L. Markert (eds.), *Developmental Toxicology: Mechanisms and Risk* 123–36. Cold Spring Harbor: CSH-laboratory.

Koch, P. B., and D. Bückmann. (1987). Hormonal control of seasonal morphs by the timing of ecdysteroid release in *Araschnia Levana* L. (Nymphalidae: Lepidoptera). *Journal of Insect Physiology* 33:823–29.

Kooijman, S. A. L. M. (1993). *Dynamic Energy Budgets in Biological Systems: Theory and Applications in Ecotoxicology*. Cambridge: Cambridge U.P.

Kuipers, T. A. F. (1986). Explanation by specification. *Logique et Analyse* 29:509–21.

Kulesa, P. M., G. C. Cruywagen, S. R. Lubkin, P. K. Maini, J. Sneyd, and J. D. Murray. (1994). Modelling the spatial patterning of the teeth primordia in the lower jaw of *Alligator mississippiensis. Proceedings 2nd Conference on Mathematics Applied to Biology and Medicine*, Lyon.

Lakoff, G. (1987). *Women, Fire and Dangerous Things*. Chicago: University of Chicago Press.

Lakoff, G., and M. Johnson. (1980). *Metaphors We Live By*. Chicago: University of Chicago Press.

Lang, J. W., H. Andrews, and R. Whitaker. (1989). Sex determination and sex ratios in *Crocodylus palustris*. *American Zoologist* 29:935–52.

Lawrence, P. A. (1992). *The Making of a Fly*. Oxford: Blackwell Scientific Publications.

Lehrman, D. S. (1970). Semantic and conceptual issues in the nature-nurture problem. In L. R. Aronson et al. (eds.), *Development and Evolution of Behavior*. San Francisco: Freeman & Co.

Levins, R. (1966). The strategy of model building in population biology. *American Scientist* 54:421–31.

Levins, R. (1968). *Evolution in Changing Environments*. Princeton: Princeton U.P.

Levins, R. (1993). A response to Orzack and Sober: Formal analysis and the fluidity of science. *The Quarterly Review of Biology* 68:547–55.

Levins, R., and R. C. Lewontin. (1985). *The Dialectical Biologist*. Cambridge (Mass): Harvard U.P.

Lewin, R. (1984). Why is development so illogical? *Science* 224:1327–29.

Lewin, R. (1993). Genes from a disappearing world. *New Scientist* (29 May): 25–29.

Lewontin, R. C. (1969). The bases of conflict in biological explanation. *Journal of the History of Biology* 2:35–46.

Lewontin, R. C. (1974a). *The Genetic Basis of Evolutionary Change*. New York: Columbia U.P.

Lewontin, R. C. (1974b). The analysis of variance and the analysis of causes. *American Journal of Human Genetics* 26:400–11. Also in R. Levins and R. C. Lewontin (eds.), *The Dialectical Biologist*. Cambridge (Mass.): Harvard U.P.

Lewontin, R. C. (1983a). The corpse in the elevator. *New York Review of Books* (January 20): 34–37.

Lewontin, R. C. (1983b). Darwin's revolution. *New York Review of Books* (June 16): 21–27.

Lewontin, R. C. (1991). Foreword. In A. I. Tauber (ed.), *Organism and the Origins of Self* ix–xix. Dordrecht: Kluwer.

Lewontin, R. C. (1992a). The dream of the human genome. *New York Review of Books* (May 28): 31–40.

Lewontin, R. C. (1992b). *Biology as Ideology: The Doctrine of DNA*. New York: Harper.

Lickliter, R. (1990). Premature visual stimulation accelerates intersensory functioning in bobwhite quail neonates. *Developmental Psychobiology* 23:15–27.

Lickliter, R., and H. Banker. (1994). Prenatal components of intersensory development in precocial birds. In D. J. Lewkowicz and R. Lickliter (eds.), *The Development of Intersensory Perception: Comparative Perspectives*. Hillsdale (N.J.): L. Erlbaum.

Lie, R. T., A. J. Wilcox, and R. Skjaerven. (1994). A population-based study of the risk of recurrence of birth defects. *The New England Journal of Medicine* 331:1–4.

Lively, C. M. (1986). Canalization versus developmental conversion in a spatially variable environment. *The American Naturalist* 128:561–72.

Longino, H. E. (1990). *Science as Social Knowledge*. Princeton: Princeton U.P.

Mackie, J. L. (1974). *The Cement of the Universe*. Oxford: Clarendon Press.

Maturana, H. R., and F. J. Varela. (1987). *The Tree of Knowledge; The Biological Roots of Human Understanding*. Boston: New Science Library.

Maynard Smith, J., R. Burian, S. Kauffman, P. Alberch, J. Campbell, B. Goodwin, R. Lande, D. Raup, and L. Wolpert. (1985). Developmental constraints and evolution. *The Quarterly Review of Biology* 60:265–87.

Mayr, E. (1961). Cause and effect in biology. *Science* 134:1501–6.

Mayr, E. (1976): Cause and effect in biology. In E. Mayr, *Evolution and the Diverisity of Life* 359–71. Cambridge (Mass.): Harvard U.P.

Mayr, E. (1993). Proximate and ultimate causations. *Biology and Philosophy* 8:93–94.

Mayr, E. (1994). Response to John Beatty. *Biology and Philosophy* 9:357–58.

McKinney, M. L., and K. J. McNamara. (1991). *Heterochrony*. New York: Plenum.

McLachlan, J. A., R. R. Newbold, C. T. Teng, and K. S. Korach. (1992). Environmental estrogens: Orphan receptors and genetic imprinting. In T. Colborn and C. Clement (eds.), *Chemically-Induced Altertions in Sexual and Functional Development: The Wldlife / Human Connection* 107–12. Princeton: Princeton Scientific Publishing Co.

Meinhardt, H. (1982). *Models of Biological Pattern Formation.* London: Academic Press.

Mestel, R. (1994). Gulls and terns all at sea about sex. *New Scientist* (3 September): 9.

Meyer, A. (1987). Phenotypic plasticity and heterochrony in *Cichlasoma managuense* (Pisces, Cichlidae) and their implications for speciation in Cichlid fishes. *Evolution* 41:1357–69.

Mill, J. S. (1973 [1843]). *A System of Logic.* Toronto: University of Toronto Press.

Mitchell, S. D. (1987). Competing units of selection?: A case of symbiosis. *Philosophy of Science* 54:351–67.

Mitchell, S. D. (1992). On pluralism and competition in evolutionary explanations. *American Zoologist* 32:135–44.

Moehrle, A., and R. Paro. (1994). Spreading the silence: Epigenetic transcriptional regulation during *Drosophila* development. *Developmental Genetics* 15:478–84.

Mousseau, T. A., and H. Dingle. (1991). Maternal effects in insect life histories. *Annual Review of Entomology* 36:511–34.

Murphy, J. P. (1990). *Pragmatism.* Boulder: Westview Press.

Murray, J. D. (1989). *Mathematical Biology.* Berlin: Springer-Verlag.

Nagel, E. (1961). *The Structure of Science.* New York: Harcourt, Brace & World.

Nanney, D. L. (1957). The role of the cytoplasm in heredity. In W. D. McElroy and B. Glass (eds.), *A Symposium of the Chemical Basis of Heredity.* Baltimore: Johns Hopkins Press.

Newman, R. A. (1989). Developmental plasticity of *Scaphiopus couchii* tadpoles in an unpredictable environment. *Ecology* 70:1775–87.

Newman, R. A. (1992). Adaptive plasticity in Amphibian metamorphosis. *BioScience* 42:671–78.

Newman, S. A. (1988). Idealist biology. *Perspectives in Biology and Medicine* 31:353–68.

Newman, S. A. (1989). Genetic engineering as metaphysics and menace. *Science and Nature* 9/10:113–24.

Newman, S. A., and W. D. Comper. (1990). "Generic" physical mechanisms of morphogenesis and pattern formation. *Development* 110:1–18.

Niimi, T., O. Yamashita, and T. Yaginuma. (1993). A cold-inducible *Bombyx* gene encoding a protein similar to mammalian sorbitol dehydrogenase; yolk nuclei-dependent gene expression in diapause eggs. *European Journal of Biochemistry* 213:1125–31.

Nijhout, H. F. (1990). Metaphors and the role of genes in development. *BioEssays* 12:441–46.

Nijhout, H. F. (1991). *The Development and Evolution of Butterfly Wing Patterns*. Washington: Smithsonian Institute Press.

Nowak, R. (1994). Moving developmental research into the clinic. *Science* 266:567–68.

Nüsslein Volhard, C. (1994). Of flies and fishes. *Science* 266:572–74.

Nylin, S. (1989). Effects of changing photoperiods in the life cycle regulation of the comma butterfly, *Polygonia c-album* (Nymphalidae). *Ecological Entomology* 14:209–18.

Orzack, S. H., and E. Sober. (1993). A critical assessment of Levins's "The strategy of model building in population biology" (1966). *The Quarterly Review of Biology* 68:533–46.

Oster, G., and P. Alberch. (1982). Evolution and bifurcation of developmental programs. *Evolution* 36:444–59.

Oster, G. F., and J. D. Murray. (1989). Pattern formation models and developmental constraints. *Journal of Experimental Zoology* 251:186–202.

Oster, G. F., N. Shubin, J. D. Murray, and P. Alberch. (1988). Evolution and morphogenetic rules: The shape of the vertebrate limb in ontogeny and phylogeny. *Evolution* 42:862–84.

Oyama, S. (1981). What does the phenocopy copy? *Psychological Reports* 48:571–81.

Oyama, S. (1985). *The Ontogeny of Information*. Cambridge (Mass.): Cambridge U.P.

Oyama, S. (1988). Stasis, development and heredity. In M. W. Ho and S. W. Fox (eds.), *Evolutionary Processes and Metaphors* 255–74. Chichester: John Wiley & Sons.

Oyama, S. (1989). Dogma: Do we need the concept of genetic programming in order to have an evolutionary perspective? In M. R. Gunnar and E. Thelen (eds.), *Systems and Development* 1–34. Hillsdale (N. J.): Lawrence Erlbaum Associates.

Oyama, S. (1992a). Is phylogeny recapitulating ontogeny? In F. J. Varela and J. P. Dupuy (eds.), *Understanding Origins* 227–32. Dordrecht: Kluwer.

Oyama, S. (1992b). Ontogeny and phylogeny: A case of metarecapitulation? In P. Griffiths (ed.), *Trees of Life* 211–39. Dordrecht: Kluwer.

Oyama, S. (1993). How shall I name thee? The construction of natural selves. *Theory and Psychology* 3:471–96 (special issue: B. Bradley and W. Kessen [eds.], Frontiers of Developmental Theory).

Oyama, S. (1994). Rethinking development. In P. K. Bock (ed.), *Handbook of Psychological Anthropology* 185–96. Westport: Greenwood.

Oyama, S. (1995). The accidental chordate: Contingency in developmental systems. *South Atlantic Quarterly*. 94(2): 509–26.

Pepper, S. C. (1961 [1942]). *World Hypotheses*. Berkeley: University of California Press.

Petersen, N. S. (1990). Effects of heat and chemical stress on development. In J. G. Scandalios (ed.), *Genomic Responses to Environmental Stress*. San Diego: Academic Press.

Peterson, K., and C. Sapienza. (1993). Imprinting the genome: Imprinted genes, imprinting genes, and a hypothesis for their interaction. *Annual Review of Genetics* 27:7–31.

Proctor, R. N. (1991). *Value-Free Science?* Cambridge (Mass.): Harvard U.P.

Radder, H. (1989). Rondom realisme. *Kennis en Methode* 13:295–314.

Raff, R. A., and T. C. Kaufman. (1991 [1983]). *Embryos, Genes and Evolution*. Bloomington: Indiana U.P.

Ransom, R. (1982). *A Handbook of Drosophila Development*. Amsterdam: Elsevier.

Redborg, K. E., and E. G. Macleod. (1983). *The Developmental Ecology of Mantispa uhleri Banks (Neuroptera: Mantispidae)*. Urbana & Chicago: University of Illinois Press.

Rendel, J. M. (1962). The relationship between gene and phenotype. *Journal of Theoretical Biology* 2:296–308.

Rescher, N. (1984). *The Limits of Science*. Berkeley: University of California Press.

Rollo, C. D. (1994). *Phenotypes*. London: Chapman & Hall.

Rollo, C. D., and D. M. Shibata. (1991). Resilience, robustness, and plasticity in a terrestrial slug, with particular reference to food quality. *Canadian Journal of Zoology* 69:978–87.

Rorty, R. (1982). *Consequences of Pragmatism*. Brighton: Harvester.

Rose, S. (1995). The rise of neurogenetic determinism. *Nature* 373:380–82.

Rose, S., R. C. Lewontin, and L. J. Kamin. (1984). *Not in Our Genes*. London: Penguin.

Ruben, D. H. (1993). Introduction. In D. H. Ruben (ed.), *Explanation* 1–16. Oxford: Oxford U.P.

Russell, E. S. (1982 [1916]). *Form and Function*. Chicago: University of Chicago Press.

Salmon, W. C. (1984). *Scientific Explanation and the Causal Structure of the World*. Princeton: Princeton U.P.

Salmon, W. C. (1989). Four decades of scientific explanation. In P. Kitcher and W. C. Salmon (eds.), *Scientific Explanation* 3–219. Minneapolis: University of Minnesota Press.

Sander, K. (1985). The role of genes in ontogenesis—evolving concepts from 1883 to 1983 as perceived by an insect embryologist. In T. J. Horder, J. A. Witkowski, and C. C. Wylie (eds.), *A History of Embryology* 363–95. Cambridge: Cambridge U.P.

Sapp, J. (1987). *Beyond the Gene*. Oxford: Oxford U.P.

Sapp, J. (1991). Concepts of organization. In S. F. Gilbert (ed.), *A Conceptual History of Modern Embryology* 229–58. New York: Plenum Press.

Scharloo, W. (1987). Constraints in selection response. In V. Loeschke (ed.), *Genetic Constraints and Adaptive Evolution*. Berlin: Springer.

Scharloo, W. (1988). Selection on morphological patterns. In G. de Jong (ed.), *Population Genetics and Evolution*. Berlin: Springer.

Scharloo, W. (1989). Developmental and physiological aspects of reaction norms. *BioScience* 39:465–71.

Scharloo, W. (1991). Canalization: Genetic and developmental aspects. *Annual Review of Ecology and Systematics* 22:65–93.

Scheiner, S. M. (1993). Plasticity as a selectable trait: Reply to Via. *The American Naturalist* 142:371–73.

Scheiner, S. M., and R. F. Lyman. (1989). The genetics of phenotypic plasticity I. Heritability. *Journal of Evolutionary Biology* 2:95–107.

Schlichting, C. D., and M. Pigliucci. (1993). Control of phenotypic plasticity via regulatory genes. *The American Naturalist* 142:366–70.

Schmalhausen, I. I. (1949). *Factors of Evolution*. Philadelphia: Blakiston.

Schmidt, K. (1994). Meet genetics' master chefs. *New Scientist* (23 April):32–35.

Sesardic, N. (1993). Heritability and causality. *Philosophy of Science* 60:396–418.

Shapiro, A. M. (1976). Seasonal polyphenism. *Evolutionary Biology* 9:259–333.

Shapiro, A. M. (1984). Experimental studies on the evolution of seasonal polyphenism. In R. A. Vane-Wright and P. R. Ackery (eds.), *The Biology of Butterflies*. London: Academic Press.

Shatz, C. J. (1992). The developing brain. *Scientific American* (September): 35–41.

Sintonen, M. (1993). In search of explanations: From why-questions to Shakespearean questions. *Philosophica* 51 (1):55–81.

Skelly, D. K., and E. E. Werner. (1990). Behavioral and life-historical responses of larval American toads to an Odonate predator. *Ecology* 71:2313–22.

Slack, J. M. W. (1991). *From Egg to Embryo* (2nd ed.). Cambridge: Cambridge U.P.

Slack, J. M. W., P. W. H. Holland, and C. F. Graham. (1993). The zootype and the phylotypic stage. *Nature* 361:490–92.

Smith, K. C. (1992a). The new problem of genetics: A response to Gifford. *Biology and Philosophy* 7:331–48.

Smith, K. C. (1992b). Neo-rationalism versus neo-Darwinism: Integrating development and evolution. *Biology and Philosophy* 7:431–51.

Smith-Gill, S. J. (1983). Developmental plasticity: Developmental conversion versus phenotypic modulation. *American Zoologist* 23:47–55.

Sober, E. (1985). *The Nature of Selection.* Cambridge (Mass.): MIT Press.

Sober, E. (1992). Screening-off and the units of selection. *Philosophy of Science* 59:142–52.

Spitze, K. (1992). Predator-mediated plasticity of prey life history and morphology: *Chaoborus Americanus* predation on *Daphnia pulex. The American Naturalist* 139:229–47.

Stearns, S. C. (1989). The evolutionary significance of phenotypic plasticity. *BioScience* 39:436–45.

Stent, G. S. (1985). Thinking in one dimension: The impact of molecular biology on development. *Cell* 40:1–2.

Sterelny, K., K. C. Smith, and M. Dickison. (1996). The extended replicator. *Biology and Philosophy* 11:377–403.

Stone, R. (1992). Swimming against the PCB Tide. *Science* 255:798–99.

Stone, R. (1994). Environmental estrogens stir debate. *Science* 265:308–10.

Stryker, M. P. (1994). Precise development from imprecise rules. *Science* 263:1244–45.

Sultan, S. E. (1987). Evolutionary implications of phenotypic plasticity. In M. K. Hecht, B. Wallace, and G. T. Prance (eds.), *Evolutionary Biology* vol 21. 127–78. New York: Plenum Press.

Suppes, P. (1957). *Introduction to Logic.* Princeton: Van Nostrand.

Suzuki, D. T., A. J. F. Griffiths, J. H. Miller, and R. C. Lewontin. (1986). *An Introduction to Genetic Analysis.* New York: W. H. Freeman.

Tarski, A. (1965 [1st ed. 1941]). *Introduction to Logic.* Oxford: Oxford U.P.

Thompson, W. D'Arcy. (1961 [1st ed. 1917]). *On Growth and Form.* Cambridge: Cambridge U.P.

Van der Steen, W. J. (1990). Interdisciplinary integration in biology? An Overview. *Acta Biotheoretica* 38:23–36.

Van der Steen, W. J. (1993a). *A Practical Philosophy for the Life Sciences*. Albany: State University of New York Press.

Van der Steen, W. J. (1993b). Towards disciplinary disintegration in biology. *Biology and Philosophy* 8:259–75.

Van der Steen, W. J. (1995). *Facts, Values, and Methodology: A New Approach to Ethics*. Amsterdam: Rodopi.

Van der Weele, C. N. (1993a). Explaining embryological development: Should integration be the goal? *Biology and Philosophy* 8:385–97.

Van der Weele, C. N. (1993b). Metaphors and the privileging of causes. *Acta Biotheoretica* 41:315–27.

Van Fraassen, B. C. (1980). *The Scientific Image*. Oxford: Oxford U.P.

Van Noordwijk, A. J., and M. Gebhardt. (1987). Reflections on the genetics of quantitative traits with continuous environmental variation. In V. Loeschke (ed.), *Genetic Constraints on Adaptive Evolution*. Berlin: Springer-Verlag.

Van Tienderen, P. H., and J. Antonovics. (1994). Constraints in evolution: On the baby and the bath water. *Functional Ecology* 8:139–40.

VanBogelen, R. A., and F. C. Neidhardt. (1990). Ribosomes as sensors of heat and cold shock in *Escherichia coli*. *Proceedings of the National Academy of Sciences USA* 87:5589–93.

Via, S. (1987). Genetic constraints on the evolution of phenotypic plasticity. In V. Loeschke (ed.), *Genetic Constraints on Adaptive Evolution*. Berlin: Springer-Verlag.

Via, S. (1993). Adaptive phenotypic plasticity: Target or by-product of selection in a variable environment? *The American Naturalist* 142:352–65.

Von Eckardt, B. (1992). *What is Cognitive Science?* Cambridge (Mass.): MIT Press.

Waddington, C. H. (1942). Canalization of development and the inheritance of acquired characteristics. *Nature* 150:563–65.

Waddington, C. H. (1947). *Organisers and Genes.* Cambridge: Cambridge U.P.

Waddington, C. H. (1953). Genetic assimilation of acquired characters. *Evolution* 7:118–26

Waddington, C. H. (1956). *Principles of Embryology.* London: Allen & Unwin.

Waddington, C. H. (1957). *The Strategy of the Genes.* London: Allen & Unwin.

Waddington, C. H. (1975). *The Evolution of an Evolutionist.* Edinburgh: Edinburgh U.P.

Wallach, M. A., and L. Wallach. (1983). *Psychology's Sanction for Selfishness.* San Francisco: W. H. Freeman & Co.

Wallach, M. A., and L. Wallach. (1990). *Rethinking Goodness.* New York: State University of New York Press.

Walsh, D., K. Li, J. Wass, A. Dolnikov, F. Zeng, L. Zhe, and M. Edwards. (1993). Heat-shock gene expression and cell cycle changes during mammalian embryonic development. *Developmental Genetics* 14:127–36.

Watson, J. D., and J. Dewey. (1987). *Molecular Biology of the Gene, vol. II: Specialized Aspects* (4th ed.). Menlo Park: Benjamin Cummings.

Webster, G., and B. C. Goodwin. (1981). History and structure in biology. *Perspectives in Biology and Medicine* 25:39–62.

Webster, G., and B. C. Goodwin. (1982). The origin of species: A structuralist approach. *Journal of Social and Biological Structure* 5:15–47.

Weider, L. J., and J. Pijanowska. (1993). Plasticity of *Daphnia* life histories in response to chemical cues from predators. *Oikos* 67:385–92.

Whitehead, A. N. (1978 [1929]). *Process and Reality.* New York: The Free Press.

Whittle, J. R. S. (1983). Litany and creed in the genetic analysis of development. In B. C. Goodwin, N. Holder, and C. C. Wylie (eds.), *Development and Evolution* 59–74. Cambridge: Cambridge U.P.

Williamson, D. I. (1992). *Larvae and Evolution: Toward a New Zoology.* New York: Chapman and Hall.

Willmer, P. (1990). *Invertebrate Relationships*. Cambridge: Cambridge U.P.

Wolpert, L. (1991). *The Triumph of the Embryo*. Oxford: Oxford U.P.

Wolpert, L. (1994). Do we understand development? *Science* 266:571–72.

Woolgar, S. (1988). *Science: The Very Idea*. Chichester: Ellis Horwood.

Coda

As promised in the foreword, in this coda I want to alert the reader briefly to some new developments and recent sources on the main themes of this book.

1. Environmental Influences in Ontogeny

As an illustration of the neglect of environmental influences in developmental biology, I mentioned (see Chapter 1) Scott Gilbert's authoritative textbook *Developmental Biology*. On the one hand I noticed that this book was special because it did at least contain various examples of environmental influences. On the other hand it illustrated their marginal position, because they were typically dealt with in special sections called "Sidelights and Speculations." When I said this at a conference in 1993, Gilbert responded that he was extremely sympathetic to my criticism, but he couldn't do anything, because where were the data he needed? However, this year the fifth edition of his textbook appeared and it contains a whole new chapter, called "Environmental regulation of animal development." This chapter provides a substantive focus on environmental influences in development, normal as well as abnormal ones. Gilbert places much emphasis on the vulnerability of the developing organism to disrupting influences (teratogens), discussing in this context the effects of retinoic acid (which is harmful when present in abnormal amounts), thalidomide, alcohol, some viruses, and environmental estrogens, concluding that most chemicals have not been screened for their teratogenic effect and that such screening is of major importance.

More generally, the exploration of environmental regulation of development is only just beginning, he says (p. 838). The chapter illustrates that data are indeed still relatively scarce and fragmented, but that they do exist and that it is important to look for them.

The environmental estrogens, which I mentioned in Chapters 5 and 6, have received increasing attention since 1995, partly due to *Our Stolen Future* by Colborn et al. (1996), which addressed the issue in urgent tones directing itself at nonbiologists and biologists alike. Colborn et al. stress that for many decades the developmental effects of chemicals went unnoticed because the right questions were not asked. Embryos were thought to be safe in their eggs and wombs, their development being guided by genes; toxicological research was directed mainly at effects in adults. New questions have become urgent. They too emphasize that the majority of chemicals have not been tested for their developmental effects. The most well-known effects (though certainly not the only ones) by now are the estrogen-like influences of a wide variety of substances. Cancer is among the late effects of such prenatal estrogen exposure. As breast cancer is known to be sensitive to estrogens, the authors urge further research here. They state that although there is now a flurry of publicity about breast cancer genes, only about 5 percent of breast cancers are the result of inherited genetic suspectibility. The study of hormone-disrupting chemicals affecting cancer deserves a higher priority than the quest for breast cancer genes, they argue, because insight into the role of environmental factors offers hope that ways can be found to prevent the disease in the majority of victims. In other words, resources should be reallocated so as to cover various causes of cancer in a more balanced way.

It is controversial how devastating the estrogen-like effects of environmental chemicals really are; some scientists think that the warning signals exaggerate their impact. For lack of data, many questions remain open, which is why Gilbert, in his chapter on environmental influences, deals with the environmental estrogens in the "Sidelights and Speculations"-section. But data on this subject are now accumulating, and they continue to point to endocrine disruption by a variety of toxic substances in a variety of organisms (see, for example, Jobling et al. 1996, and Gerritsen 1997).

2. Integrating Evolution and Development

The integration of evolution and development is increasingly taking shape. In a fascinating review called "Resynthesizing Evolutionary and Developmental Biology," Gilbert et al. see evidence that three classical embryological themes that had been eclipsed by the genetic redefinition of evolution and development, are rediscovered in new forms. These themes are macroevolution, homology and, above all, the morphogenetic field, which they see as the central integrating concept.

Now Gilbert et al.'s view is exactly that: a view, not the last word. The rediscovered fields they describe, for example, are defined in terms of information from genes and gene products that becomes translated into spatial entities. Such fields differ from those of the embryologists of the first half of this century and they also differ Goodwin's fields that were discussed in Chapter 2 of this book. So this proposal will not end all discussion and controversy. It does illustrate, however, how strong the synthesizing forces have become. In this process, development is gaining so much importance for evolution that the authors feel confident that population genetics (the central integrating field of the modern synthesis) will have to change if it is not to become irrelevant to evolutionary theory.

Though synthesis may be the trend, the different tenets of the three approaches distinguished in Chapter 2 have not disappeared. This could be illustrated through an analysis of present field-concepts, through pointing at continued gene-centric research as well as continued complaints about it, or through the continuing disagreement about the role of history in biological explanation. Let me add a few words about this last theme. Goodwin's view that biological explanation of form should not be historical (see Chapter 2) has been attacked by Paul Griffiths (1996), who writes from the perspective of "Developmental Systems Theory" (the constructionist approach). In aiming to explain generic forms, he says, structuralists implausibly assume that most unrealized forms are not viable. But on the assumption that the vast majority of the viable forms in morphospace are not realized (which he finds more plausible, given what is known about the Cambrian explosion of forms), the explanation of why some forms, rather than other ones, are realized should focus on the contingencies of history: "The idea of generic form can and should be historicized."

The process of integrating evolution and development thus remains a complex and multivoiced one. Among the voices is also Jablonka and Lamb's book *Epigenetic Inheritance and Evolution* (1995). It makes a strong plea for the importance of environmental influences and for "a wider view of heredity." These authors have written before about the phenomenon of epigenetic inheritance, for example through methylation patterns; I referred to their earlier work in Chapter 4. In the book, they give an overview of what is known about epigenetic inheritance and directed mutation, and they argue that a "Lamarckian dimension" should be recognized, not in opposition to neo-Darwinism, but as an addition to it. In their view, the assumptions underlying DNA-centric research programs should be changed, and more research should be carried out that is based on the insight that the environment influences the epigenetic inheritance system. For example, "the possibility that environmentally induced epigenetic diseases can be transmitted through several generations will have to be incorporated into epidemiological research programmes" (p. 275).

3. Conclusion

A gene-centered approach to living organisms has yielded the view that knowing everything about DNA is the Holy Grail of biology. In her book *Refiguring life*, Evelyn Fox Keller (1995) remarks that a funny thing happened on the way to this Holy Grail. Information yielded by new techniques for analyzing genes and their behavior has subverted the doctrine of the gene as the sole (or even primary) agent of development (p. 22). Reality appears to be much more complex. Some relevant new findings on this complexity are reviewed by Eva-Neumann-Held (1996). Among them is the recently discovered phenomenon of mRNA-editing, which involves the insertion, removal or conversion of nucleotides in mRNA molecules. As Neumann-Held stresses, this phenomenon illustrates the reciprocal contingency of the components of developmental systems. Yet at the same time Fox Keller warns us that the models we now have are only a beginning in the study of interactive and emergent phenomena and that there is no guarantee this promising line will continue. Tools are important elements in determining directions, conceptual as well as technical tools. I remain convinced that the concept of reaction norm is an important and powerful one. Jessica Bolker (1995) has called attention to another tool: the model systems used in developmental biology. In particular, she states that the favorite model systems have been selected for their flat reaction norms, that is, for their relative insensitiviy to environmental influences. Little is detected about environmental influences through such model systems, and as a consequence, an overly deterministic view of development has taken hold.

All in all, two years later, work on environmental estrogens and other developmental toxigens as well as work on epigenetic inheritance and gene regulation reinforces the conclusions of Chapters 5 and 6. Developmental processes should be approached as interactive and ecological phenomena, and special emphasis on environmental influences in the developmental network is important for reasons of health and survival.

Bolker, Jessica. (1995): Model Systems in Developmental Biology. In *BioEssays* 17:451-55.

Colborn, Theo, J. P. Meyers, and D. Dumanoski. (1996): *Our stolen future.* Boston: Little, Brown and Co.

Gerritsen, A. (1997): *The influence of body size, life stage, and sex on the toxicity of alkylphenols to* Daphnia magna. Thesis, Utrecht University.

Gilbert, Scott F. (1997): *Developmental Biology* Fifth edition. Sunderland (Mass.): Sinauer Associates.

Gilbert, Scott F., J. M. Opitz, and R. A. Raff. (1996): Review: Resynthesizing evolutionary and developmental biology. *Developmental Biology* 173: 357-72.

Griffiths, Paul. (1996): Darwinism, process structuralism and natural kinds. *Philosophy of Science* 63 (Proceedings): S1-S9.

Jablonka, Eva, and M. J. Lamb. (1995): *Epigenetic Inheritance and Evolution; The Lamarckian Dimension.* Oxford: Oxford U.P.

Jobling, S., D. Sheahan, J. A. Osborne, P. Matthiessen, and J. Sumpter. (1996): Inhibition of testicular growth in rainbow trout (*Onchorhynchus mykiss*) exposed to estrogenic alkylphenolic chemicals. *Environ. Toxicol. Chem.* 15: 194-202

Keller, Evelyn Fox. (1995): *Refiguring Life; Metaphors of Twentieth-Century Biology.* New York: Columbia U.P.

Neumann-Held, E. M. (1996): The gene is dead—Long live the gene: conceptualising the gene in a constructionist way. In P. Koslowski (ed.): *Developmental Systems, Competition and Cooperation in Sociobiology and Economics.* Berlin: Springer-Verlag

Index